THE FUTURE AFTER AI

Expectations for Artificial Intelligence as the New Operating System of Finance, Technology, Energy, Healthcare, Education, and Business

Jason Schenker

THE FUTURE AFTER AI

Expectations for Artificial Intelligence as the New Operating System of Finance, Technology, Energy, Healthcare, Education, and Business

BY JASON SCHENKER

ISBN: 978-1-957772-40-0 *Kindle*
978-1-957772-37-0 *Paperback*
978-1-957772-24-0 *Hardback*

PRESTIGE
PROFESSIONAL PUBLISHING

For my wife, Ashley.

CONTENTS

CONTENTS

CONTENTS

PREFACE

THE FUTURE AFTER AI

One conversation can change your life. It can be the foundation of a professional origin story, the origin of a publishing and research mission, and the cornerstone of a book — just like this one. In late September 2016, I had one such conversation with Martin Buehler, who was then the head of robotics and R&D at Disney Imagineering.

For my entire professional career, my professional mission has been to serve corporate clients by providing them with economic research and consulting support. That was true in my first role as an investment bank trading desk economist at Wachovia (now Wells Fargo) in 2004. It remained true when I worked in consulting at McKinsey, and it became the central mission of the boutique financial market research firm, Prestige Economics, which I founded in 2009.

Beginning in 2016, a growing number of Prestige Economics corporate clients were seeking a broader understanding about the future of automation, robotics, and work. So, my mission took me to the RoboBusiness conference, where I met Martin.

The RoboBusiness conference included a number of robots on display, speakers on the future of automation, and networking opportunities. In a break, I struck up a conversation with Martin. A roboticist with a Ph.D. from Yale, Martin is as brilliant as he is thoughtful. He shared a genuine enthusiasm for my interest and the goals of my research trip. I was trying to uncover what the future would look like, and Martin was very kind to share his thoughts with me as a newbie at RoboBusiness.

It was just the two of us sitting at a round conference table, as he told me that the biggest challenge in the world of robotics was how long it takes to get a roboticist through education to a point where they could actually design and make robots. He explained a painfully long academic process, where high school graduates go to college, grad school, and then do post-doctoral work before they can really work in the space.

In other words, a roboticist might not be ready to enter the workforce and make a contribution until almost age 30. Martin shared stories and strategies that were being implemented in the technology world to get roboticists up and running with just a couple of years of community college.

"When we've got that done, all of the problems for making more robots will be solved," he said. I thought about it for a second and asked, "Is anyone working on the labor market impacts when we have unlimited robots?"

After a thoughtful pause, he answered, "No."

It was that moment that changed my career.

In that moment, I realized the world was racing toward a future no one was preparing workers or leaders to navigate. I thought "someone needs to do this," and, realizing that there wasn't anyone else to do it, I decided to help people prepare for emerging technological challenges.

Just a few days after returning from RoboBusiness, I founded The Futurist Institute with a mission to help professionals and leaders prepare for the coming changes and impacts of technology, managing the downside risks and capturing the upside opportunities. Founding The Futurist Institute supported multiple major research and training initiatives.

Research from my attendance at RoboBusiness was incorporated into the publication of my bestselling book, *Jobs for Robots*, in February 2017. My next course of action to help leaders prepare for the future was to create an online training program, including courses, a certification, and a professional designation, which is the Certified Futurist and Long-Term Analyst or FLTA®.

The Futurist Institute also created *The Robot and Automation Almanac*, which included annual editions from 2018, 2019, 2020, 2021, and 2022. While the title did not explicitly call out artificial intelligence, AI was incorporated into numerous almanacs countless times.

Even OpenAI — the nonprofit that launched ChatGPT in November 2022 — was directly referenced in the 2019 edition of the almanac, which was published in December 2018. I also incorporated detailed discussions of OpenAI's work with large language models in my talks in early 2019.

All of these experiences, which began with my pivotal conversation with Martin, helped me prepare countless professionals and leaders for the recent surge in AI adoption. The Futurist Institute FLTA® certificate holders, Prestige Economics clients, and the various organizations, associations, and entities I advise were well-positioned for the global push into AI that accelerated when ChatGPT launched in November 2022.

Being prepared for AI also led to new global opportunities, including when Seoul Mayor Oh Se-hoon appointed me to a three-year term as the AI and Emerging Technology Advisor to the Seoul International Policy Advisory Group in September 2025. The experience underscores a powerful lesson of this book: invest early in understanding transformative technologies so you can put yourself on the path of future opportunities.

Acknowledgements

First, I want to thank **Martin Buehler** for taking the time to set me on the path of helping others to prepare for the future.

I want to thank my friend and former colleague, **Nawfal Patel**, who worked with me to help build The Futurist Institute, edit and publish editions of *The Robot and Automation Almanac*, and support the mission to help professionals prepare for the future impacts of technology on economic and labor market conditions.

A special thanks to **Kerry Ellis** for her exceptional work on the cover. I wanted a cover that paired old and new, while also providing a visual cue to the idea that artificial intelligence will be the bedrock of the future economy. She created an amazing cover.

Most importantly, I want to thank my family for their unwavering support. Writing a book is always a demanding process, and my family has supported me in countless ways — offering both emotional support and editorial feedback.

Every book-writing journey spills over into family life, and for their patience, understanding, and encouragement, I cannot thank them enough. I am especially grateful to my loving wife, **Ashley Schenker**, whose encouragement has meant everything.

I also want to express my deep appreciation for my wonderful parents, **Janet Schenker** and **Jeffrey Schenker.** When I founded The Futurist Institute in 2016 and wrote *Jobs for Robots* in 2017, both of them provided feedback and shared their thoughts on my research and mission. Sadly, in 2021, my father passed away, and my mother moved into a memory care unit that same year. Though they will never see how far technology has come, I am forever grateful for their love, wisdom, and guidance.

Finally, thank you for reading *The Future After AI*.

My goal is to offer insights that empower you to successfully navigate the future. As you explore the pages ahead, I hope this content helps you prepare for the future, managing downside risks and seizing upside opportunities in the years ahead and beyond.

~ Jason Schenker
November 2025

INTRODUCTION

WHY I WROTE THIS BOOK

Artificial intelligence is impacting countless industries, and while there is a debate about whether its future is being overhyped or underhyped, one thing seems certain: AI will be the bedrock of the future economy.

AI will be like combustion engines, indoor plumbing, electricity, movable type, the telephone, cell phones, personal computers, and the internet. Although all of these technologies count as great wonders that changed the world, we no longer marvel at their existence because they are assumed. And yet, if we were to try to compare the lives of our ancestors, these technologies would be some of the most critical developments in terms of lifestyle, worker output, productivity, and communication.

We haven't stopped talking about these technological advancements because they are irrelevant. We have stopped talking about these marvels, because we take them for granted as cornerstones of our modern existence.

That is why I wrote *The Future After AI* — to provide the insights, futurist frameworks, and strategic perspective necessary to help leaders from industry, the civilian government, and national security entities understand and prepare for this new era of global competition.

In the book ahead, there is a discussion of how AI is becoming the foundational technology of the future economy, how people can strengthen the uniquely human skills that set them apart, how major industries are being reshaped by AI, and how leaders can use futurist strategies to navigate the economic, geopolitical, and operational challenges ahead.

The first section of this book, **AI Opportunities**, examines how artificial intelligence is quickly shifting from something new and surprising into something foundational and largely taken for granted. Just like we eventually stopped talking about electricity, semiconductors, or the internet because they became assumed parts of life, we are on the cusp of doing the same with AI.

In this section, I walk through the early signs of this shift: rising productivity, labor market pressures that incentivize AI adoption, macroeconomic imperatives to adopt AI, and the practical value of generative and agentic AI across day-to-day work. I also highlight some of the risks and guardrails organizations need to consider as they adopt these tools.

When leaders understand what AI can do right now and what is coming soon, they can capture the upside faster and avoid the pitfalls that will hurt the companies and individuals who fall behind.

The second section of this book, **Rising to Meet AI Challenges**, delves into the human side of the AI story. As AI handles more mid-office and back-office tasks, the real differentiators become distinctly human: presence, trust, judgment, relationships, and originality. These differentiators are the domain of the front-office, client-facing, customer-facing, and people-facing roles.

This section focuses on how people can stand out by doubling down on what only humans can do. It includes recommendations about building relationships, identifying your highest-value work, understanding your professional worth, and bringing human originality to create work that AI can't compete with. In a world with powerful AI tools, your personal connection with others and the value you create that cannot be automated will matter more than ever.

The third section of this book, **Impacts After AI**, explores how AI is reshaping major sectors of the economy. I walk through the future of finance, technology, energy, healthcare, education, and business, showing how AI is already driving big changes and where it is likely to take us next. You will see how AI could transform finance and audit, push global energy demand to new highs, support healthcare workers, challenge traditional education models, and reshape how businesses reach customers.

This section also introduces my expectations for B2AI (business-to-artificial-intelligence) sales and marketing strategies that drive AI discoverability to boost revenue. With the near-term ubiquity of AI agents, B2A (business-to-agent) sales and market strategies are the next logical step, since your next customer may not be a human at all, but an AI agent acting on their behalf.

These chapters break down sector by sector what the future after AI actually looks like in practice. There are tremendous opportunities across industries and sectors in the future after AI. But there are also some risks, including the prospects of endless growth in the number of factor authentications.

I conclude this section with a discussion of the future of cities after AI, drawing on my experience as the AI and emerging technology advisor to the Seoul International Policy Advisory Group and key insights from my keynote to the Metropolis 40th Anniversary Seoul Congress in September 2025.

The book concludes with strategic recommendations in the fourth and final section, **Strategy for the Future After AI**. Here, I bring together the futurist frameworks, geopolitical insights, and strategic tools that help leaders navigate the years ahead.

This section examines how leaders can use applied futurism, scenario planning, and structured foresight as they prepare for uncertainty in business, in national security, and across the broader economy. There is also an in-depth examination of how AI proliferation is poised to shape corporate governance mandates, Cold War Two conflict risks, and the increased importance of information missions of special operations forces.

Ultimately, this last section reinforces one of the core message of this book: the future after AI belongs to leaders and organizations that prepare early, act thoughtfully, manage risks wisely, and seize the opportunities at hand.

To help leaders navigate the opportunities and risks of AI to their organizations, businesses, and careers, I share the PAR framework that calls for organizational best practices of focusing on preparedness, adaptability, and resilience. By reinforcing these organizational qualities, leaders across industries are poised to limit their biggest risks and win in a future of unimaginable productivity benefits, wealth, and growth.

Navigating The Future After AI with Clarity and Strategy

Artificial intelligence is the reality shaping our technological, economic, and business landscapes. As this book illustrates, the defining elements of AI present myriad risks and opportunities in an increasingly complex world, requiring intentional, thoughtful, and futurist strategic approaches.

The technological, business, economic, and military risks of this era are unfolding in real time, with consequences that will reverberate for decades.

Leaders need more than historical reflection to recognize that AI will be the foundation of the future economy and the invisible operating system that drives industries, businesses, and the world in which we live. This is where futurist frameworks and strategies can support and inform forward-looking analysis, adaptive strategic planning, and decisive action.

Balancing current risks and opportunities is harder than ever. But it is easier now than it ever will be in the future.

Now, let's get to it!

SECTION 1

AI OPPORTUNITIES AND RISKS

CHAPTER 1

THE BEDROCK OF THE FUTURE ECONOMY

The big idea of this book is that in the not-too-distant future, we will stop talking about artificial intelligence. I don't expect this because AI is going away. In fact, it's quite the opposite.

I expect people will stop talking about AI because AI will be the bedrock of the future economy. AI will be understood to be a cornerstone of modern technology, alongside computers, the internet, cell phones, semiconductors, and microprocessors.

Whether people fully grasp how AI works under the hood will be largely irrelevant. The critical reality is that AI will function as a baseline component of how the world operates. It will be deeply embedded in the digital, technological, and economic operating systems of our time. That is the future *after* AI — not a world where AI is gone, but a world where AI is everywhere, essential, and almost taken for granted.

We don't talk much about semiconductors, microprocessors, or internal combustion engines in everyday conversation.

They are everywhere, yet invisible. They quietly power our devices, cars, networks, and infrastructure. Most people cannot explain how a microchip works or how a combustion engine functions in detail — and it doesn't matter. What matters is that they know what these technologies *do* and what they enable.

In the same way, I believe we are heading toward a world of *AI invisibility*. AI will be everywhere, embedded in and integrated with tools, platforms, services, and workflows. But we may not talk about it very much. We'll talk about the applications, the outcomes, the businesses, and the experiences — just not the underlying AI itself.

Google Trends Data

As a way to gauge historical interest in key innovations of the technological era, Google Trends data and graphs tell an important story that could very well play out for the future of artificial intelligence and even technologies that have not yet been fully deployed, like quantum computing.

Interest in the Internet, the Computer, and the Cell Phone

Google Trends data reflects the relative interest in web searches since 2004. A reading of 100 percent represents the all-time high interest. Plus, falling trends indicate declining web searches and interest, whereas rising trends indicate growing interest.

Worldwide interest, as measured by Google web searches, offers insights into how remarkable technology seems — and how unremarkable some technologies now appear to be. Searches for the word "internet" have fallen from a peak level of 100 percent in January 2004 to just 32 percent in October 2025.[1]

EXHIBIT 1-1[2,3]

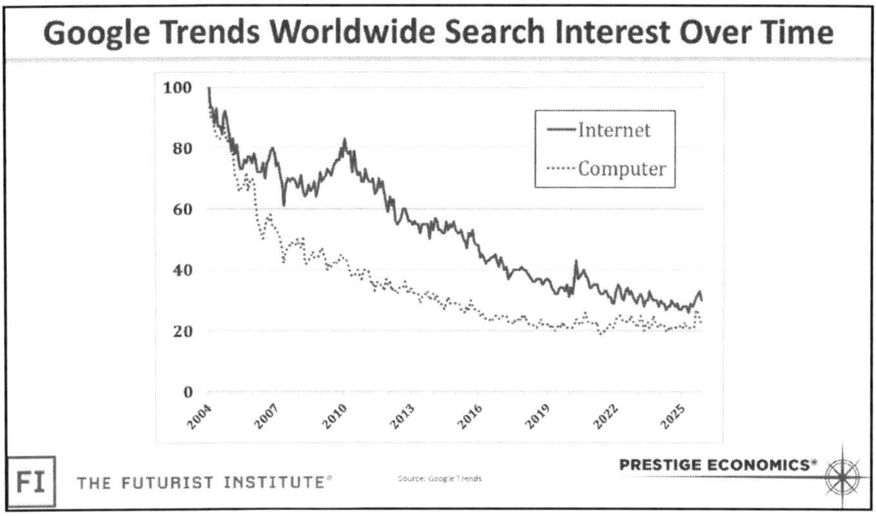

EXHIBIT 1-2[4]

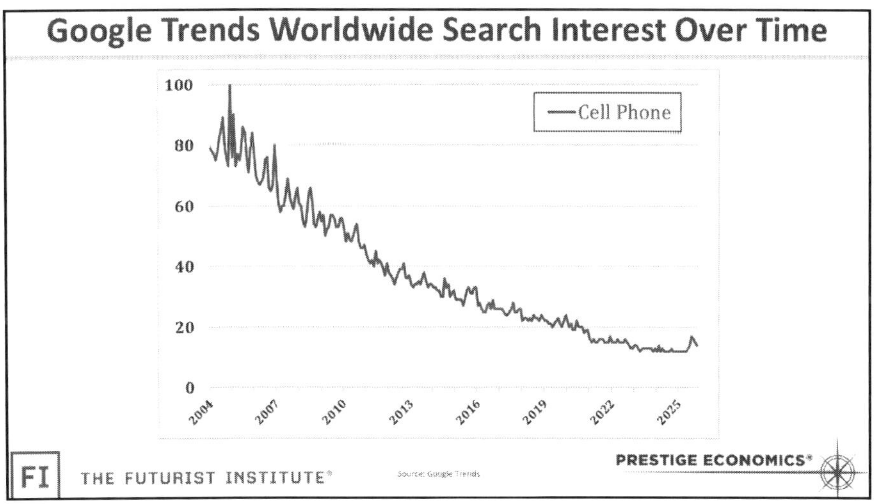

Similar trends can be seen in Google web searches for the word "computer," which have fallen from 100 percent in January 2004 to only 26 percent in October 2025.[5] Google searches for "cell phone" have dropped even more precipitously from a peak of 100 percent in 2004 to 16 percent in October 2025.[6]

Clearly, the internet, the computer, and the cell phone are now unremarkable. And this is likely to be the future of AI as well.

Past and Future Interest in AI

Worldwide Google web searches for "AI" were at only 2 percent of their peak in January 2024, but had risen to 100 percent in September 2025 — a record high peak.[7] This series reflects massive interest growth, especially following the rollout of ChatGPT in late 2022.

But all trends and interests peak at some point, and even this series fell in October 2025 to 87 percent from its September peak of 100 percent. As AI becomes foundational for the future economy, it will be taken for granted, and there will be less interest in understanding what it is and how it works.

Future Interest in Quantum Computing

One technology that is not likely to reflect anything close to its peak interest is quantum computing. Google web searches for "quantum computing" were at 100 percent in October 2025 — a record high.[8] While this series reflects a massive growth of interest, this technology has yet to be fully developed and rolled out into daily considerations. However, once this technology reaches scale, it, too, is likely to see a future decline in interest.

EXHIBIT 1-3[9]

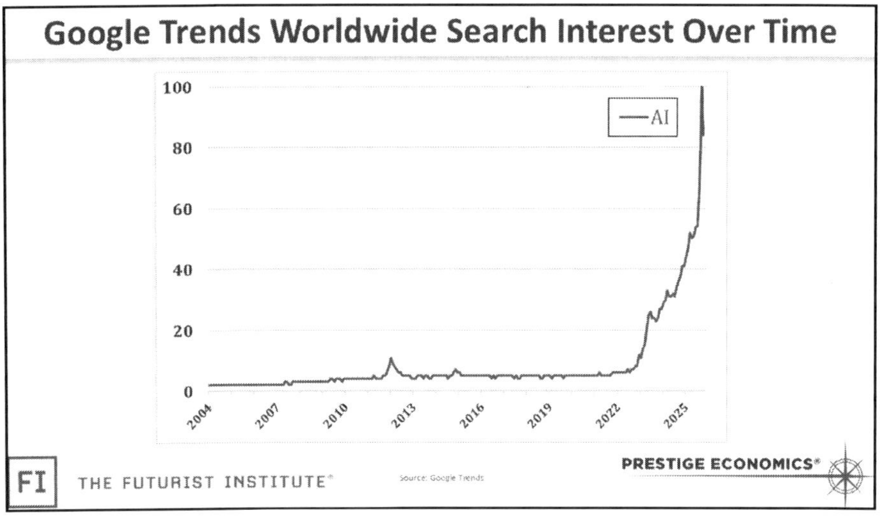

EXHIBIT 1-4[10]

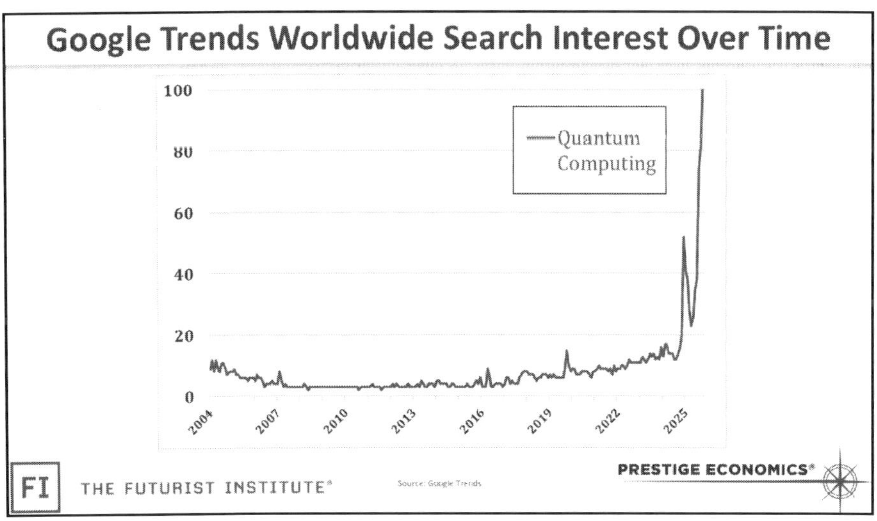

The AI-Invisible Future

In countless keynote speeches and board presentations over the past three years, people have asked me if we are in an AI bubble or hype cycle. Some technologies have been overblown in recent years — like blockchain, which I called out in my book *The Promise of Blockchain: Hope and Hype for an Emerging Technology*.

But AI does not seem to fit the bill for overhype. In my own work, I have seen the tremendous value-add of AI as a productivity lever in mid- and back-office support functions that are not core to my role.

Not all technologies will be critical for future industries and the economy, as the computer, the internet, and cell phones have been. However, artificial intelligence seems likely to be on the list of these economy-altering technologies. AI is poised to become so crucial that, eventually, the economy will become so AI-integrated and AI-invisible that search interests and discourse on this topic will fade.

CHAPTER 2

LESSONS FROM SPREADSHEETS

One of the most common economist mantras is that *this time is never different*. It's a perspective that has informed my analysis at Prestige Economics and many of the technology futurist scenarios developed by The Futurist Institute over the years. I also think this mantra applies to artificial intelligence.

When I think about the future of AI, I look to the past for indications of how the patterns of resistance and adoption are likely to play out in the future. One of the most apropos examples comes from the world of finance.

AI and Lotus 1-2-3

My father and grandfather were Certified Public Accountants who ran a small business together throughout my childhood and for most of my life. In the 1980s, I remember when Lotus 1-2-3 was introduced. It completely changed how people in finance thought about technology. Before then, doing everything by hand on paper was the norm. The idea of using a digital spreadsheet for accounting and financial analysis was new, strange, and — to many people — deeply uncomfortable.

When digital spreadsheets were rolled out, people in finance functions at the time generally fell into two camps.

One group was eager to adopt the new technology and use it as a productivity lever — to do more work, boost output, help more clients, be more efficient, and contribute to company growth.

The other group resisted. They didn't want to use the new software. They dismissed digital spreadsheets and wanted to keep doing everything by hand.

Today, it's almost impossible to imagine that debate.

If you talk to anyone working in finance now, the notion that people once *preferred* paper ledgers over digital spreadsheets sounds absurd. But it's true. And once Lotus 1-2-3 rolled out, it was eventually overtaken by Microsoft Excel, which became — and remains — the standard.

With split perspectives on the future of using digital spreadsheets, accountants and finance professionals either adopted this technology quickly, as my father and grandfather did, or they resisted it.

Those who refused to adapt were either pushed out of their jobs or their firms went out of business. Some people and companies eventually came around, adopting the very tools they had once rejected — but often only after realizing that the cost of resisting was too high. As you might expect, late adopters also risked falling behind competitors, struggling to keep up.

Implications for Artificial Intelligence

This story about a technology we now take for granted has powerful implications for the future of artificial intelligence.

Today, just a little more than three years after the launch of ChatGPT, many companies are still debating whether they should use AI, how they should use it, and where it fits in their business operations. Of course, other organizations have moved quickly to fold in AI technologies directly into their operations.

Organizations mired in endless debate about using AI are poised to struggle, and they could face the fate of accounting firms that equivocated over the value of digital spreadsheets in the 1980s, because many companies have already moved forward with AI.

Forward-looking companies that are adopting AI technologies are experimenting, building use cases, and integrating AI into processes to drive productivity, increase worker output, and reduce costs. The results of those efforts — higher productivity, lower costs, and better performance — are precisely the kinds of outcomes that historically separate the winners from the losers when a seismic technology shift is underway.

And yet, some business leaders are still debating AI.

I suspect that ten or twenty years from now, we will look back on this era the same way finance professionals in 2025 look back at the pre-spreadsheet world. Just as it seems unbelievable today that anyone once debated whether to use digital spreadsheets, it will likely soon seem incredible that anyone debated whether or not to use AI.

Implications for AI Companies

Today, in finance, it's assumed that everyone uses digital spreadsheets. I haven't met anyone in this century who does all of their finances by hand. That world is gone. And in the next few years, I think it will become just as difficult to find professionals who do not use AI in some form.

But here's a big takeaway for AI companies: no one in finance uses Lotus 1-2-3 anymore.

Lotus was the first to market and it defined the space. But Excel came to dominate the world of digital spreadsheets, and Lotus 1-2-3 lost the race. The implication for AI businesses is clear: *being first does not guarantee survival.* Just because you launch a breakthrough technology, including something like ChatGPT, does not mean you will automatically exist in the future.

Tech companies are scrambling to build their AI capabilities because they want a piece of the digital delegation gold rush — and they want to avoid being left in the dust!

And while Lotus lost out to Excel decades ago, it wasn't officially discontinued by IBM until 2014.[1] This reality hints at a potential future where some technologies might limp along for what seems like forever. The Google Trends searches for Lotus 1-2-3 were a paltry 3 percent in October 2025, but people were still searching for it.[2]

That's what keeps AI executives up at night: the idea that people might still be searching for ChatGPT 40 years from now, even if it loses the generative AI race pretty early on.

Future Opportunities

The productivity benefits of AI are too significant to ignore, but the winners have not yet been decided — and the race is on.

The potential for productivity gains at the macroeconomic level, across the entire economy, at the business level, and even at the level of individual contributors is enormous. Ignoring AI's potential — or actively resisting it — could prove catastrophic. Resisting the use of AI, or even delaying the implementation of AI technologies, could become a going-concern issue for companies and a career-ending decision for individuals.

AI will be integrated into the economy. It won't just be the "new technology." It will become embodied as a core part of the operating system of our labor force, just as the internet, computers, mobile phones, and automobiles have. These are no longer things we marvel at; they are simply part of life. And AI is likely to be viewed the same way in the future.

EXHIBIT 2-1[3]

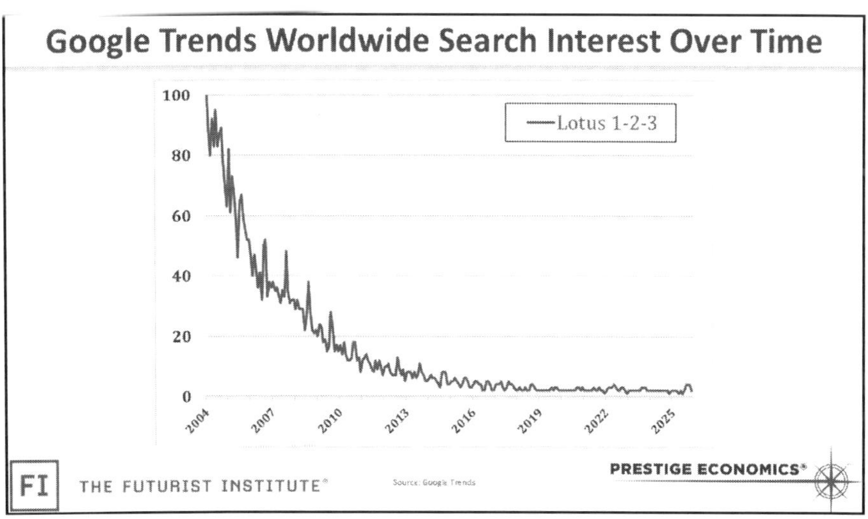

CHAPTER 3

GREAT AND NOT-SO-GREAT EXPECTATIONS

There's been a debate brewing as to whether AI is overhyped or underhyped. From my perspective, the answer is that AI may be *both* overhyped *and* underhyped at the same time.

A June 2025 Pew Research survey of Americans captured this hype divide: 21 percent said AI has been made a *bigger* deal than it really is, but 36 percent said AI has been made a *smaller* deal than it really is. In the middle, 36 percent said AI has been described about right. Five percent heard nothing about AI.[1]

EXHIBIT 3-1[2]

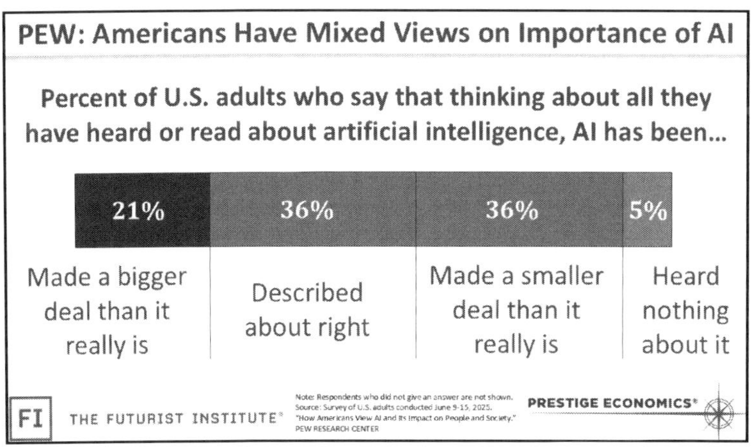

While Americans are clearly divided in their expectations of AI, the hype-versus-hope debate about AI's potential is likely to play out in the world of data. The AI value-add will hinge on how people use AI, how they treat the inputs, and how seriously they take the data.

Overhyped and Underhyped

When I was in graduate school studying applied economics about twenty-five years ago, I had a professor who ran econometric models that would take three full days to complete. These were statistical models that use some of the same math, proofs, and computational analysis as the ones that underpin today's AI, machine learning, and large language models.

To make sure the computer was never touched, she put chairs around it to keep people away. After three days, her results would appear, and that was as good as it got 25 years ago. Fortunately, today, that analysis can be done in three seconds, which presents tremendous productivity opportunities. After all, when complicated analysis shifts from three days to three seconds, it certainly hints that AI potential may be underhyped.

But the overhyped risks stem from what gets overlooked.

After all, what mattered even more in the modeling process was what happened before my professor ever hit "run." That's because she spent about three weeks cleaning and preparing the data. She verified that every column and row was correct, that every currency field was accurate, that the units of measure were correct, that there were no missing or misplaced values, and that nothing had been shifted or misaligned.

Taking the time to prepare and clean your data diligently are two best practices for data that every analyst should follow.

Today, I suspect that the three-day model would run on an AI model in three seconds. Yet I strongly suspect that in lieu of spending three weeks preparing and cleaning data, many people now spend zero seconds preparing their data before feeding it into AI systems. Skipping over data best practices can be costly.

In all of the various economic, corporate risk, and financial analyses I've conducted in my career, cleaning and preparing data correctly is a top priority. This is actually an even more crucial imperative with the adoption of artificial intelligence.

No matter what kind of AI you are using, whether quantitative models, large language models, or qualitative analysis, there is always the fundamental truth that the quality of your outputs is entirely dependent on the quality of your inputs. If the data is garbage, the results will be garbage in, garbage out. And if those results inform business strategy, financial planning, or corporate decision-making, there can be really serious negative financial and operational consequences.

Seven Pillars of Good Data Analysis

When I speak to corporations, industry associations, government groups, or even U.S. defense and intelligence entities, I always emphasize that good analysis requires a process, and I share The Futurist Institute's proprietary seven pillars of good data analysis. It begins by asking the right question, scoping the project, collecting the data, cleaning the data, analyzing the data, testing the results, and retesting the results.

There is a risk with ubiquitous AI capabilities that people will grab whatever data they have, put it into an AI tool, run it, and accept whatever comes out. There would be no scoping, no cleaning, no checking, no testing. It's not really analysis at all. It is merely an impatient Id-driven desire for an instantaneous answer, regardless of whether that answer is correct or reliable.

Some people think just hitting return to use AI to analyze the data is enough. It is not — and it creates massive risks.

There is a German term that captures this perfectly, *gefährliches Halbwissen*. It means dangerous half-knowledge. In English, the closest idea is a half-baked idea. This is the way to describe people who think they know what they are doing, know just enough to be dangerous, but lack the understanding to avoid creating significant risks.

EXHIBIT 3-2[3]

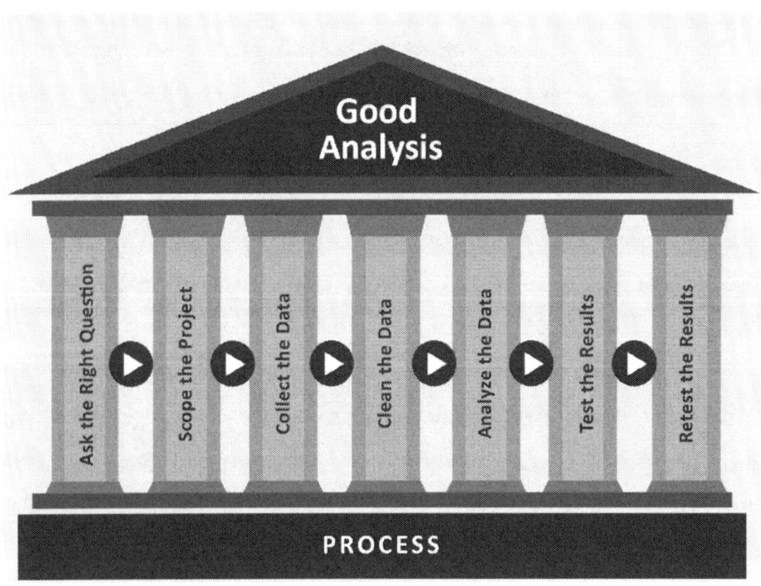

Dangerous Half-Knowledge

Making expensive mistakes due to a lack of expertise has always been a risk. It's why most major consulting firms engage experts with industry-specific knowledge when they are working on a large project, including strategy expansions, mergers and acquisitions, or asset dispositions and divestitures. Before AI, subject matter experts (SMEs) were a critical hedge against these kinds of risks. In the future, if companies and analysts rely too much on AI and don't consult SMEs before making critical decisions, risks could quickly get very expensive.

The most costly error like this I saw was about a decade ago. An energy valuation project that went horribly wrong when a private equity shop in Connecticut bought a natural gas asset in Texas. The analysts supporting the fund buying the asset did not understand the incredibly valuable distinction between dry gas (methane) and wet gas (propane, ethane, and butane). This is not a mistake anyone who has worked in oil and gas would make. But a PE fund without industry, sector, or regional experience was caught unawares — and paid a hefty price.

The result was a massive overvaluation of the asset, due to a double counting of wet gas molecules as both wet and dry gas. In other words, a multimillion-dollar mistake was made because the analysts did not understand the fundamentals and did not check with any industry experts. After the flawed, overpriced acquisition, an energy investment bank client of mine was brought in to revalue the asset properly.

Generative AI democratizes the analytical horsepower previously limited to private equity or investment bank bullpens for anyone.

That horsepower can be a double-edged sword, because AI can create the illusion of analytical competence — but it may suffer from critical fundamental information gaps that the user did not even know how to query. For those users who are not cautious, errors due to a lack of understanding could be costly.

To minimize the potential downside of unseen risks, subject matter experts (SMEs) are likely to become more valuable to keep in the loop — especially for the highest-value activities and priorities. However, there are risks that, over time, AI users' caution could experience significant attrition, raising the potential for expensive mistakes.

No Universal Answer

The question of whether AI is overhyped or underhyped does not have a universal answer. But it is intrinsically linked with the question about a culture of data.

For AI users who do not know or care about data fidelity or industry fundamentals, and only want quick results without following any process, their shortcuts will eventually come with a high cost. I am reminded of the old automotive repair adage that between cheap, fast, and good, you can only pick two. For those who pick fast and cheap, AI will not be very good — and could turn out to be quite costly.

For companies and professionals who understand data, maintain a culture of accuracy, value high fidelity, embody due care, and apply disciplined analysis, AI is underhyped. These people will gain enormous value from its speed and power.

CHAPTER 4

GENERATIVE AI OPPORTUNITIES

Generative AI presents instant productivity and value for organizations right now. In talks I give to executives, corporate boards, and industry associations, I always share the extraordinary range of generative AI use cases that can be deployed by businesses today to boost output, increase productivity, and drive financial benefits. These opportunities are immediately actionable and capable of driving measurable gains in workforce productivity across almost every mid-office and back-office function in modern businesses.

When talking to leaders, I usually frame generative AI functions like a *junior analyst* that you would delegate work to. As of late 2025, this digital junior analyst is comparable to someone with dozens of advanced master's degrees, but it's still their first day on the job. This is why it is so important for generative AI outputs to be reviewed with a healthy dose of professional skepticism and expertise. Still, the productivity leverage that generative AI already offers today is enormous. Below are some of the most valuable categories of generative AI deliverables that I discuss in my talks, because they can yield immediate benefits.

Job Postings

One of the most immediate areas of transformation is creating job postings. Human resources teams have traditionally spent hours crafting the right mix of responsibilities, qualifications, years of experience, education requirements, and cultural expectations. Generative AI can produce a high-quality draft almost instantly by drawing on historical examples, industry standards, and internal templates. HR professionals can then refine the details and focus more on strategic hiring needs instead of repetitive documentation work.

Sales Scripts

Sales teams also benefit tremendously. Developing a polished sales script used to take weeks or months of brainstorming, testing, and rewriting. AI can now generate first-round script ideas in seconds. These drafts often include customer personas, key value propositions, objection-handling language, and relevant industry terminology. With human refinement and customer testing, these drafts become strong and consistent tools that improve the performance of entire sales organizations.

RFPs

Procurement and operations teams gain efficiency from AI as well. Requests for Proposals (RFPs) take considerable time to assemble because they contain detailed technical requirements, legal language, evaluation criteria, and scope definitions. Generative AI can rapidly create a structured, comprehensive draft by using large repositories of historical content and pattern recognition. Specialists should still finalize the technical and legal sections, but AI eliminates the need to start from a blank document.

LOPs

Responses to RFPs, often presented as Letters of Proposal, are equally labor intensive. Vendors describe their capabilities, track record, team experience, pricing, terms, and timelines. AI can create a coherent and well-organized structure for these proposals, leaving human experts to fill in sensitive details such as pricing, delivery schedules, and custom project requirements. This dramatically reduces the time between receiving an RFP and submitting a competitive response.

SOPs and Training Materials

Generative AI offers powerful potential for drafting standard operating procedures and training materials. These documents require precision, clarity, compliance awareness, and consistency. In the past, they often took multiple departments weeks or months to develop, update, and circulate. But generative AI can use historical SOPs, industry guidance, and company documentation to produce near-complete drafts almost instantly. As with other generative AI outputs, business leaders and subject matter experts would still need to refine the content to ensure accuracy and alignment with internal standards.

Legal Drafts

Using generative AI to create legal *drafts* is another significant opportunity. Generative AI can draft contracts, agreements, and legal memos by drawing on widely available patterns in case law and contract language. But I stress in all my talks that it is important to continue using lawyers to review final outputs. Seriously, using lawyers is critical. But the initial stages of contracting and negotiation can be accelerated by leveraging generative artificial intelligence to get rough legal drafts in place.

These drafts are often close to correct, but in legal work the difference between *close* and *correct* can be enormously costly. AI can assist by providing a starting point that accelerates discussion and reduces drafting cycles, but all legal documents should be reviewed and approved by qualified attorneys.

Form Letters

Corporate communications teams are also finding value in AI-generated form letters and other standardized messaging. These communications often rely on established patterns that are easy for AI to replicate. Whether the letter is intended for shareholders, regulators, business partners, or customers, AI can create a draft that reflects the appropriate tone and structure of past communications.

Press Releases

Press releases are another area of immediate benefit. Press announcements require a very specific, clear structure, including headlines, quotes, factual summaries, and corporate boilerplate language. Generating a polished release used to require hours of careful drafting. Now, AI can create strong drafts in seconds, allowing communications teams to quickly edit, finalize, and distribute time-sensitive announcements.

Project Plans

Project plans are similarly transformed. A full project plan can include objectives, milestones, resource requirements, risk assessments, and work breakdown structures. These documents often take significant time to draft and organize. When I worked in consulting, a comprehensive project plan would take multiple days to craft. Today, it can be done in seconds.

While SMEs should absolutely review project plans, Generative AI can rapidly create a detailed plan framework, giving project managers a strong starting point and enabling them to focus on tailoring the plan to the organization's needs rather than building it from scratch.

Emails and Emotional Modulation

The most valuable use I have found for generative AI is in crafting emails to convey the right tone and messaging. Communications often require careful wording, emotional balance, and a thoughtful tone. Striking the right balance requires a great deal of care — and sometimes a lot of time. This is especially true for emails related to sales cycles, billing, collections, negotiations, or customer service.

The thing I hate most about running a business is collections, and having generative AI draft correspondence in these situations is a huge relief — and it allows me to focus on higher value-add activities like writing books like this one or *Cold War Two*.

For my business, email modulation is hands down the most tremendous value-add and time-saving offering of generative AI.

It is so effective because AI draws on an enormous repository of writings that help it generate drafts that are clear, professional, and aligned with the intended tone of the message. There is even the ability to tweak the tone and messaging to be firmer or softer, which can greatly reduce communication risk and save considerable time for managers and executives.

Productivity Gains

Across these areas, AI has the potential to deliver meaningful productivity gains by eliminating the blank-page problem and accelerating the drafting process for a wide range of deliverables. But organizations need to remain attentive to oversight and should keep experts in the loop to actively refine, edit, and approve key deliverables — especially those that will be public-facing.

Work tied directly to revenue, legal commitments, or financial exposure should *always* be reviewed by human experts. AI may be fast and highly capable, but it still makes errors and lacks the judgment required in high-stakes decision-making.

A Digital Swiss Army Knife

Used appropriately, generative AI is not a replacement for human professionals. Instead, it can be a digital Swiss Army Knife that can be a force multiplier that improves efficiency, expands capacity, and frees people to focus on higher-value work. The organizations that learn to integrate AI as a reliable drafting partner will be better positioned to grow quickly, operate efficiently, and compete effectively in the years ahead.

CHAPTER 5

LABOR MARKET NEEDS

The most essential ingredient for the success of AI is demand, which is why the tight U.S. labor market and a rise in the cost of labor are fueling an AI boom.

Despite commentary about the weakening U.S. labor market in the second half of 2025, the job market has remained extremely tight — especially for physically demanding in-person roles. Plus, in the 24 months between September 2023 and September 2025, payrolls expanded in all but two months.

Despite the rise in the number of net U.S. jobs, the unemployment rate rose from 3.8 percent to 4.4 percent. But there are two caveats. First, even though the unemployment rate has risen, 4.4 percent is an exceptionally low unemployment rate and equal to the lowest rates seen between June 2001 and May 2018.[1] Plus, the rise in unemployment rate between September 2023 and September 2025 was not the result of widespread job losses.

The unemployment rate increased because more people are entering the labor force than there are jobs being created.

In other words, job creation continued — but was not fast enough to keep the unemployment rate from rising.

Labor Force Participation and Unemployment Claims

In September 2025, the labor force reached a new record high of more than 171.2 million people — but it still isn't enough people to meet all of the U.S. labor market needs. On the same date that the September jobs data were released, the weekly jobless claims report showed that less than 2 million people were collecting unemployment insurance, which is roughly 1.1 percent of the labor force, and one of the lowest percentages in history. In other words, even though the September 2025 unemployment rate was 4.4 percent, the number of people actually collecting unemployment benefits reflected that the job market is even tighter than the unemployment rate reflects.[2]

To put the broader health of the labor market in perspective, there are also over 7.2 million open jobs in August, which is 5.2 million more than the number of people collecting unemployment. Even though openings have fallen from the March 2022 peak of 12.2 million, they remain far above the levels seen during the 2001 recession and the 2007-2009 recessions, when openings were only between 4 and 5 million.[3]

Where the Jobs Are

The distribution of job openings is critical. Recent college graduates have seen their unemployment rate rise to 4.8 percent[4], reflecting both high interest rates and the impact of AI on entry-level roles. But acyclical hiring in sectors like healthcare remains strong, accounting for the lion's share of monthly payrolls in recent months.

EXHIBIT 5-1[5]

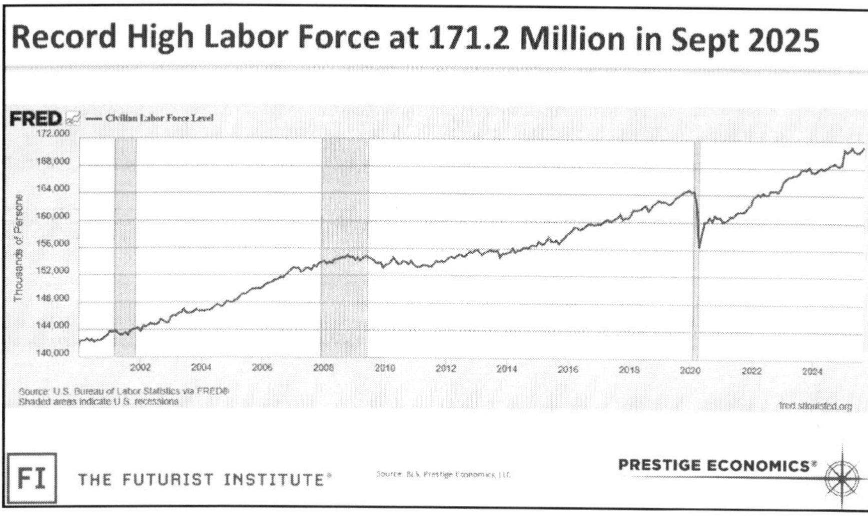

EXHIBIT 5-2[6]

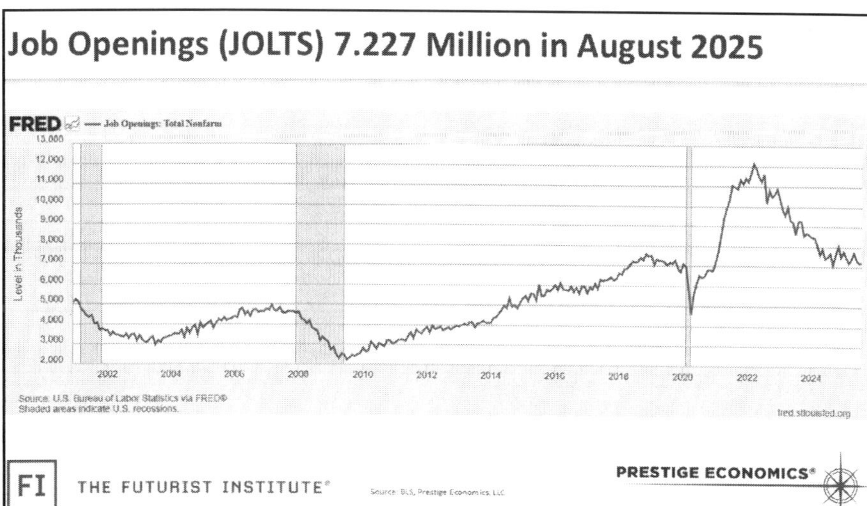

Healthcare and social assistance job openings alone accounted for nearly 1.4 million open jobs in August[7], which represents around 70 percent of the people collecting unemployment benefits nationwide. This sector, along with other physically demanding, in-person roles, has become the primary driver of labor demand. In contrast, entry-level professional services roles face weaker demand.

Entry-Level Work, AI, and Hiring Disincentives

AI has become a direct substitute for many entry-level research, administrative, and analytical tasks. Simple research projects that once required interns or junior analysts can now be handled by tools like ChatGPT, Copilot, or Gemini. While these tools require judgment and oversight, they arrive with more advanced capabilities than many recent graduates. This is likely a key reason why the unemployment rate for recent college graduates (ages 22-27) has risen to 4.8 percent in June 2025.[8]

AI is not perfect. It is similar to a graduate with 50 master's degrees who has never worked a day in their life. It can produce valuable work, but should be supervised with healthy skepticism. Still, its existence reduces the economic justification for hiring new graduates, especially in a high-interest-rate environment.

As AI tools remain extremely inexpensive, far cheaper than hiring staff, businesses are becoming increasingly reliant on them. This mirrors the early days of Uber. It was initially cheap and subsidized, and later became much more expensive once demand was established. AI subscriptions may eventually rise dramatically, but even at higher prices, they will still often replace entry-level labor.

EXHIBIT 5-3[9]

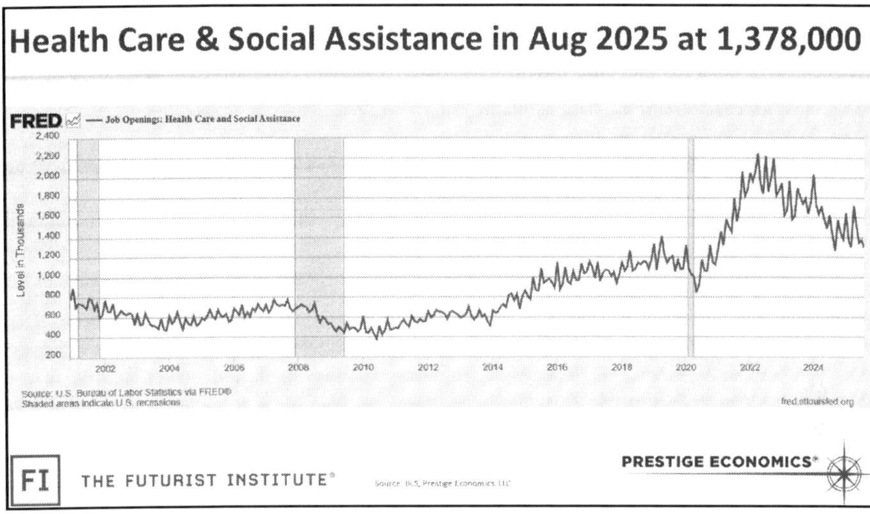

EXHIBIT 5-4[10]

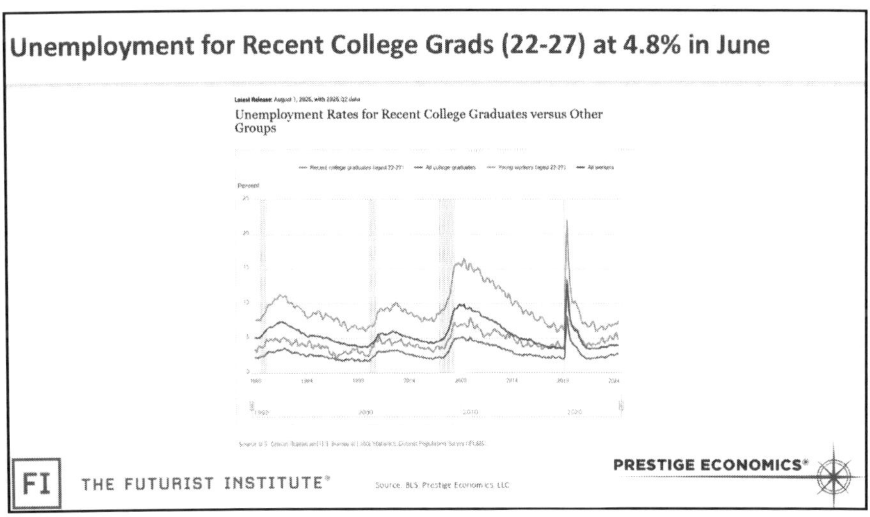

Rising Demand in the Physical Economy

While AI reduces demand for many entry-level office jobs, demand for physically demanding, in-person roles is very strong. Healthcare and social assistance are the largest and most persistent sources of job growth. Meanwhile, in August 2025, the manufacturing sector had more than 400,000 open jobs, far above pre-Great Recession levels. Plus, transportation, warehousing, and utilities show roughly 355,000 openings, triple the levels seen before the 2001 and 2007 recessions.[11]

As the United States prepares for a wartime-ready industrial base, physical-economy jobs will become even more important. Industrial production, shipbuilding, metals, infrastructure, and supply chain resilience all require expanded labor.

Wages have risen sharply, up by $8.24 per hour between January 2020 and September 2025.[12] This is a massive increase and has been contributing to inflationary pressures. In September 2025, core services inflation was up 3.5 percent year on year, driven largely by domestic costs such as healthcare and shelter.[13] These inflationary pressures are only likely to ease by solving some of the labor shortages we see in the JOLTS data.

Wages for Physically Demanding In-Person Roles (PDIRs)

The demand for PDIRs is likely to remain high and continue to rise further as demographic trends drive demand in healthcare. At the same time, reindustrialization and national priorities are poised to drive even greater labor demand in the manufacturing, warehousing, transport, and utilities sectors. These pressures present a potential for bifurcated future wage dynamics across the U.S. economy.

EXHIBIT 5-5[14]

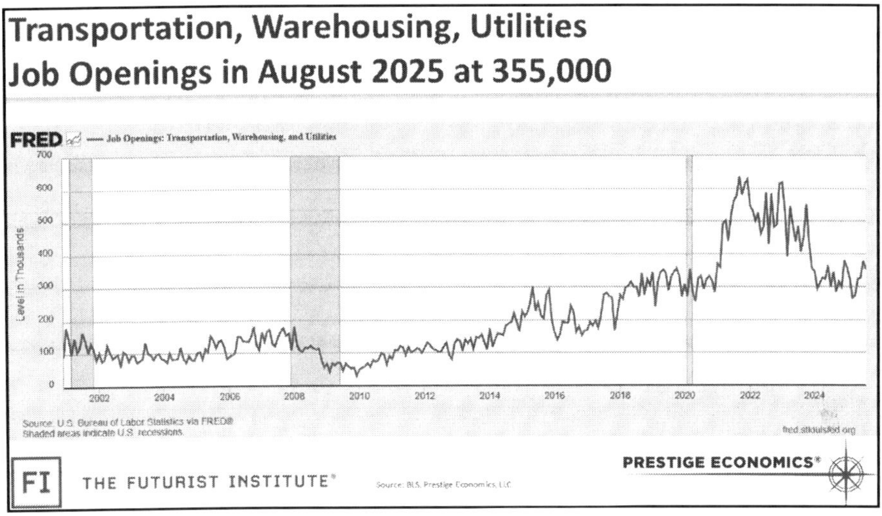

EXHIBIT 5-6[15]

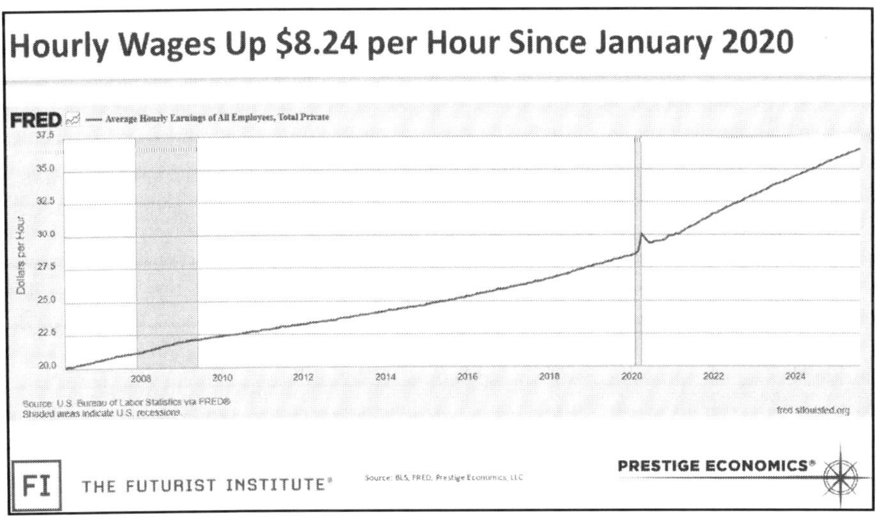

On the one hand, jobs in professional and business services are likely to experience flat wages or even wage compression and layoffs as companies use AI to boost productivity and stretch resources. At the same time, automation and robotics solutions for PDIRs remain unaffordable or impractical, which is likely to result in higher wages to fill those labor market needs. This means that the inflationary pressures on U.S. wages are likely to be mixed.

As a financial professional, one thing I have learned in my career is that all markets clear. By that, I mean that supply and demand come together at some price point. We are not yet at that price in healthcare and other sectors, and while AI may keep some professional wages under pressure, the unmet labor market needs are likely to be resolved with higher wages. It is an inelegant solution, but it is the most likely.

Conclusion

The U.S. labor market is tight, and wages have risen significantly in recent years, providing fundamentally strong incentives for companies to deploy AI and automated solutions wherever possible. The massive unmet labor market needs for PIDRs are likely to push those wages higher. Meanwhile, companies will be under pressure to maintain profit margins, which is why business leaders will be highly incentivized to embrace AI as a means to boost company output and mitigate rises in wages across industries and sectors other than PIDRs.

CHAPTER 6

MACROECONOMIC OPPORTUNITIES

Macroeconomic opportunities presented by artificial intelligence cannot be understated, because AI represents one of the greatest labor force productivity levers in economic history. It also presents a payroll tax hedge for businesses and corporations.

AI Productivity Potential

Consider that the U.S. economy had almost 160 million people on payrolls as of September 2025.[1] If that workforce becomes merely five percent more productive due to AI tools, that would be the same full-time equivalent (FTE) output of eight million additional workers that essentially appeared overnight. But the upside does not stop there. As AI becomes integrated into enterprise systems, these gains compound.

Five percent is a very modest near-term productivity boost that we might experience across the labor force in the near term. For me, AI has boosted my productivity by around 30 percent. If we see a 30 percent productivity boost across the economy in the next decade, that would be the equivalent of adding 48 million FTE worker output into the U.S. economy.

In a future where highly personalized AI systems could become almost complete virtual digital twins that can operate alongside every professional, productivity gains could reach 70 percent or more. That would be the equivalent of over 100 million additional workers' worth of intellectual output to the U.S. economy. And that is just in the United States. The global impact would be so much larger.

Understandably, some investors worry about the potential for overbuilding and a bubble. Valuations are high and can shift rapidly due to a variety of factors. One thing that seems true is that AI presents massive economic productivity potential and is likely to be foundational for the future economy and the new operating system for countless industries.

AI Use Cases are Clear

AI is not like past technologies that were misunderstood at launch. Anyone who has used a generative AI program for even a few minutes can see firsthand the potential productivity boost it offers.

Compare that to the famous TV interview Bill Gates gave to David Letterman following the launch of Windows 95, Letterman poked holes in the value and use cases for the personal computer, joking that you could already use other technologies like the radio and tape recorders that perform the same functions. The audience laughed heartily.[2]

With AI, the use cases are immediate, obvious, and economically powerful. Everyone understands what it can do — and that it will do more, which explains the surge in global AI investment.

U.S. Entitlements Risks Incentivize AI and Automation

One of the core arguments I made in my 2017 book, *Jobs for Robots*, is that long-term fiscal risks loom over the U.S. financial system. AI and automation effectively act as corporate hedges against future tax increases tied to entitlement funding shortfalls.

Rising entitlements costs, including Social Security, Medicare, and Medicaid, have contributed to a rising U.S. national debt that has surpassed $38 trillion.[3] Moreover, those entitlement programs are teetering on the verge of insolvency.

Increases in life expectancy, falling birth rates, and shifting population pyramids point to a future further decline in the ratio of workers per Social Security recipient. These demographic shifts almost guarantee higher payroll taxes in the future, presenting growing long-term labor cost risks to employers.

The future trajectory of the United States' entitlements programs is creating a powerful incentive for companies to accelerate the adoption of AI, automation, and robotics because entitlements are funded by payroll taxes, which are split between employers and employees.

Social Security insolvency is projected for the mid-2030s. Medicare had long been forecasted to face insolvency as early as 2026. However, Medicare appears unlikely to become insolvent in 2026 — and it is not expected to have solvency issues until some time in the 2030s. The reason is straightforward: payroll taxes were increased. This dynamic offers a clear preview of what is likely to happen again: when entitlement programs face future funding crises, policymakers are likely to raise payroll taxes.

As demographic and financial strains on entitlements programs intensify, the likelihood of rising payroll taxes increases, with real implications for employer cost structures. The lack of payroll tax liability on automated systems reinforces employer incentives to replace human labor with automated solutions.

AI Prices to Rise with a Potential Shift to Tokenization

The promise of AI offers massive productivity potential for the macroeconomy at the population level. Plus, AI and automation also offer the potential for businesses to minimize their future payroll tax liabilities. This means employers will increasingly find more reasons to adopt AI, robotics, and other automated tools.

Robots do not pay payroll taxes — at least not yet. But at some point, policymakers may look at automated systems and consider imposing payroll-type taxes on AI and robotics to support entitlements programs. That day may come, but it is not here yet. For now, employers see rising risks associated with human labor costs and the opportunity to use AI to reduce headcount-related expenses and potential future tax burdens.

With the potential to bolster growth, lower costs, and offer businesses a hedge against future taxes, AI is the trifecta of business opportunity. That's why so much investment is going into AI. It also means AI offerings are way too cheap — for now.

With so much potential, watch out for rising prices! The future cost structure for AI seems likely to involve tokenization, with use charges that extract maximum value from users. But even at much higher prices, the ROI and productivity value-adds of AI for businesses and the economy are likely to be massive.

AI tokenization pricing could allow for more effective price discrimination, whereby AI users with a higher propensity to pay (usually due to a higher perceived ROI) will pay more. These are the same economic theories that underpin flight pricing models, where passengers pay different prices for the same service.

In the end, more price discrimination would enable AI companies to realize more revenues as they extract more value from more customers — especially from the customers who have the highest propensity and ability to pay. Additionally, if AI programs are able to implement use-based pricing, we could see a massive shift from subscription-based software-as-a-service (SaaS) sales models to token use models across multiple areas of technology.

Conclusion

The productivity potential of artificial intelligence is likely to be massive, and entitlements programs' tax funding methods are also likely to encourage companies to embrace AI. Moreover, with such high-value propositions for businesses, companies will likely pay more —and potentially a lot more — for AI services in the future.

To manage the potentially endless demand for AI services as well as boost revenue and profitability, AI companies are likely to shift to tokenization pricing models that embody more perfect price discrimination, extracting higher revenues from customers who expect the highest ROI from using AI. Finally, if tokenization pricing replaces SaaS pricing for AI, this could spread to other technology industries, driving revenues, profits, shareholder valuc, and valuations even higher for some tech companies.

CHAPTER 7

AI RISKS AND GUARDRAILS

Artificial intelligence presents significant risks and limitations that fall into several buckets. These include operational, social, and confidentiality. Knowing the risks upfront is critical for using AI effectively to capture opportunities while limiting downside risks.

Operational Risks

The biggest host of risks for businesses and companies using AI includes operational risks that could adversely impact how a company operates and performs.

The most significant operational risk is if people become overly dependent on AI, leading to a proliferation of errors as they lose the ability to distinguish between quality work and flawed deliverables. In a similar vein, it is vital to ensure that workers are able to do their jobs *without* AI. That means companies need to train people in a way that builds real capability first. Only then can AI become a tool that enhances work rather than replacing underlying skills.

For example, I have written hundreds of press releases over the years. And yes, in recent years, I've used generative AI to help draft most of them. But when I use ChatGPT to create a press release, I feel confident because I know what I'm looking for. I know what could be wrong. I know what might be missing. I know how a press release *should* read. That confidence came from years of doing it myself without AI.

In addition to making sure people can do a job without AI first, companies should also provide ongoing learning and testing opportunities to maintain initial and ongoing credit quality. The goal is to ensure that technology makes workers more effective, instead of just making them AI-dependent.

AI can help improve worker deliverables. But without strong human oversight, the quality will slip. In the future, executive function, editing judgment, and human review will matter more than ever.

Social Risks

AI also presents significant social risks, including bias, dullness, hallucinations, and accuracy drift.

Because AI draws on historical data, it may embody and reinforce the same past biases that exist in that data. People have often wondered what policing in a future with Robocops might look like. The answer will depend significantly on how historical policing data about crime, arrests, and incarcerations are controlled for bias and incorporated into the new operating systems.

Another social risk is that everything becomes dull. By late 2024, more than half of global online content was estimated to be AI-generated. And if AI keeps pulling from the past, recombining what already exists, and regurgitating it over and over again, the world could lose a lot of originality and creativity.

A third key social risk lies with hallucinations. Sometimes AI programs can generate content that does not exist or is not true. Although this has improved over the past two years, I have seen this firsthand. Early versions of ChatGPT indicated that I have degrees I do not have from schools I attended, and that I had worked at places I have never worked. It also mentioned a book I had written, except that *neither I nor anyone else on earth* had written such a book.

There is one final social risk that AI presents — and it bleeds into corporate risks as well. This is the issue of content and accuracy drift, whereby AI could actually get worse over time. That can happen unintentionally, or through adversarial activity. An AI can take in more information and somehow become less accurate. More data doesn't always mean better results.

Confidentiality Risks

The third AI big bucket of risk is confidentiality — and this is a big one.

There is a real risk when people put proprietary or sensitive corporate or personal information into generative AI systems. This includes computer code, legal contract terms, draft trademark or patent filings, client information, trade secrets, or anything confidential.

A simple rule I share with every client is this:
If you would not want it to appear on the front page of The *New York Times* or in the inbox of your competitors, do not put it into a generative AI program.

Your data is not like an anchovy on a salad that you can simply remove. It's like the paprika in a goulash. Once it's in, it's everywhere. There is no going back.

Ignoring confidentiality risks can have massive costs for organizations and even be career-ending for individuals.

While social risks are tough to control for, confidentiality risks can be reduced and mitigated. My top recommendation to clients, executives, and associations is to limit what you share with AI by creating and sharing clear rules about what can and cannot be submitted to AI systems.

AI Risks and Guardrails

AI Guardrails help you get the benefits of AI without falling into the traps. Writ large, there needs to be a cultural norm of caution when it comes to using AI.

It is crucial to ensure that people can do their work without AI assistance, that they try to control for biases in historical data, and that there are strict protocols to prevent any sensitive or proprietary data from ever ending up in a generative system.

If something is too sensitive for public view, it is too sensitive for an AI system.

AI outputs should also be viewed with skepticism. As in other elements of work and life, it's a good idea to treat AI with a strategy of *trust, but verify*.

Adopting these kinds of guardrails and guidelines can help you to minimize some of the core risks of using AI.

Conclusion

AI is powerful. It is transformative. But it also has limits. Understanding these limits is the key to using AI wisely. Because winners in the future will be the people who combine strong human skills with the strategic use of intelligent tools, if you build those skills and put sufficient guardrails in place, AI will make your work better. If you don't, the risks will grow.

CHAPTER 8

AGENTIC AI RISKS AND LIMITATIONS

One of the AI fields with the greatest potential is agentic AI, in which AI acts as your agent to perform tasks for you. Even in an era of rapid technological advancement, there will be important limits on where, when, and how AI can be used effectively. And while there are big opportunities, there will also be significant risks and limitations.

The Limits of Agentic AI

Working with an AI agent raises fundamental questions about trust, oversight, and accountability. Anyone who has worked with a bad real estate agent, travel agent, or financial adviser understands this challenge immediately: agents do not always do the job you want them to do. They make sloppy mistakes, miss opportunities, take risks, or otherwise act in a way you would not want them to.

In other words, even though they represent you, they may not represent you well.

And the same could happen with AI agents.

Most of the time, even human agents require oversight. Because the agent represents you, their mistakes can quickly become your consequences.

The same will be true of agentic AI.

A professor friend of mine addressed this dynamic perfectly in the early days of ChatGPT by allowing his students to use AI when writing research papers — but with the understanding that he would grade their work entirely on accuracy. If the AI hallucinated or made an error, the student was held responsible. In other words, if the AI made a mistake, the student's grade was marked down.

The tension in this situation, and an innovative way to incorporate the risks of unsatisfactory AI outcomes and behaviors, offer a preview of the future of agentic AI: if your AI agent acts on your behalf, you will still bear the repercussions.

These risks explain why agentic AI may require a human to *be in the loop*, and involved in every step of the process — or at least *on the loop*, supervising the process but not approving every action.

Why Trust Is Difficult to Delegate

Travel is a quintessential example that tech thought leaders invoke when discussing the future use cases for agentic AI. But personally, I think this use case is wildly overhyped — and presents significant risks that people have not considered.

In a client meeting in November 2025, a client asked me about the potential for AI travel agents. Without hesitation, I stated that I would never outsource my travel planning, because it is a core part of my business. It's why I don't already outsource my travel, nor have I ever outsourced travel bookings.

If I don't trust a travel agent or secretary to bring me the best options, why would I trust an AI system?

Despite how this may sound, I am not a Luddite. Rather, I simply recognize the importance of maintaining control over the outcomes of a core competency.

After two decades of traveling as a speaker, taking hundreds of flights a year, arranging my travel has become one of my distinctive core competencies. I know the airports, the risks, and the schedules better than most human agents. Most importantly, I know the consequences of poorly picking the wrong flight, the wrong airline, the wrong flight connection time, the wrong route, or even the wrong seat.

In 21 years, I have never missed an event.

When things have gone awry, as they sometimes have, I have changed up my travel to avoid hurricanes, canceled flights, and long delays. Since I book my own travel, I know that if something goes wrong, I only have myself to blame. And it is up to me to make it right.

Consider a scenario in which an AI agent booked a tight connection based solely on historical data. The agent might make that booking without understanding the realities of a specific airport or airline's track record. The agent may also have ignored that it is hurricane season or other weather-related risk factors. But most importantly, that AI agent will never know the experience of sprinting to a boarding gate only to watch the door close right in front of its eyes.

I would be the one forced to live with that mistake, and I would be the one running for the boarding gate, thinking the whole time that my AI is an idiot. In essence, I would be doing what the computer told me to, not the other way around.

For me, travel is executive judgment and a core competency of my business. It's not the kind of thing that should be outsourced.

If I only traveled occasionally, perhaps I would feel differently. But for frequent travelers who make their money by efficiency and effectively traversing countless airports every year, the stakes are too high.

Travel is mission-critical to my job, so it's non-delegable.

When an AI Agent Fails

All of this talk of travel and agentic AI begs one question: who is responsible when an AI agent fails? This question applies far beyond travel. Would you let a human agent you've never spoken to buy or sell assets for you? Negotiate contracts? Commit you to obligations?

Millions of human agents do these things, including real estate agents, business brokers, travel agents, lawyers, and financial fiduciaries. Chances are, you've worked with people who do these tasks. But you probably spoke to them first. And you know that their work comes with three essential elements: reputation, accountability, and a "throat to choke."

If you wouldn't let a human do these things, would you let an AI agent? Or if you would only let a human do these things for you with careful consideration, how likely are you to let agentic AI do it for you?

Plus, if a human agent fails, there is someone to answer to. If a digital agent fails, the consequences fall entirely on you.

Legal work is a key example. Drafts created by AI can be helpful. But allowing an AI system to create a full legal contract without attorney review is extremely risky and outright dangerous. If the contract contains errors, you might only realize this in a courtroom when you are surrounded by real attorneys at a cost that is infinitely higher than if you had hired an attorney to begin with. One attorney I know likes to say that good attorneys are expensive, bad attorneys cost way more in the long run.

Agentic AI Use Cases Require Oversight

The automation of many activities presents opportunities, and there are many potentially value-add use cases for agentic AI. But if you wouldn't let a secretary do these things now, don't plan on letting agentic AI to do them anytime soon either. Consider a few potential use cases often cited for agentic AI: sending emails, managing workflows, and ordering gifts.

First, let's consider letting an AI agent send emails. It sounds really risky. I recommend trusting current AI programs as much as I would trust someone with lots of education who does not have significant work experience. That means drafts are fine. But sending emails without approval is not. Drafting emails and meeting plans is a much lower-risk activity than sending emails without review.

Second, consider letting AI agents manage workflows and project management timelines. I have to admit that imagining teams of human workers being told by an AI how they need to spend their time sounds unmitigatedly dystopian. Monitoring workflows is a significantly lower-risk task than committing people to specific activities and timelines without human review.

Third, is the idea of letting an AI agent buy people gifts. While this may sound innocuous, it presents personal risks — even if it does not present the same kinds of business risks as using AI agents to send emails or manage workflows.

Shopping for gifts is one of the most fun things you can do for someone else. You are thinking of them and how to make them happy. Why would you outsource that? I had a relative who asked his secretary to buy a holiday gift for my sister. It was a beautiful necklace, but when she found out his secretary had picked it out, my sister was furious.

Plus, there are some big downside risks. Imagine that you have a friend who has recently experienced the death of a child. It would be somewhere between insensitive and traumatic if your AI agent automatically sent children's gifts for the holidays.

If these scenarios play out, who are you going to blame when an impolite email, an unreasonable workflow request, or an insensitive gift is sent? Are you going to say it was the AI agent, further depersonalizing these actions, further eroding trust and relationships?

That's the core risk of agentic AI. It could damage or devalue the most valuable asset you have in a world of AI: relationships with other people.

Conclusion

Agentic AI will certainly be widely adopted, and many people will find it helpful. But others will be deeply frustrated when it fails in ways they did not anticipate and cannot easily correct. People will use AI more than they should, simply because the technology is cheap and accessible. But that ease of use also creates risk.

Using an AI agent to assist humans offers significant benefits. But letting agentic AI act as a principal, committing to transactions that carry legal, financial, or logistical responsibility, creates a far deeper set of risks.

At the outset, people and organizations should limit their consideration of potential agentic AI use cases that present relatively low financial, reputational, and well-being risk.

The technology will improve, but the fundamental issue remains: an agent cannot experience the consequences of its actions. Only a human can.

SECTION 2

RISING TO MEET AI CHALLENGES

CHAPTER 9

THE PERSONAL PROFESSIONAL IMPERATIVE

Artificial intelligence promises a level of analytical capability that previous generations could not have imagined. Mid- and back-office functions — analytics, research, document drafting, workflow automation, administrative tasks — are becoming faster, cheaper, and infinitely more scalable. We are entering an era when professionals will have what feels like unlimited analytical horsepower at their fingertips, a kind of digital superintelligence augmenting every process behind the scenes.

But as AI accelerates work behind the curtain, something very different is happening on the front of the stage. The more technical and operational workloads are automated, the more critical personal presence becomes. Client-facing roles will depend not on how much data someone can process, but on their ability to connect with another human being — authentically, meaningfully, and repeatedly. In the world after AI, the most impactful value will not come from what people know, but from who they know, how they connect, and how well they build relationships that technology cannot replicate.

A June 2025 Pew Research survey showed that half of Americans believe AI will make it harder for people to form meaningful relationships.[1] That finding reveals a deeper truth: as technology grows more capable, many people grow less comfortable with basic human interaction. Texting replaces calling. Notifications replace conversation. Digital proximity replaces genuine presence. And when AI can provide flawless analysis in seconds, the personal becomes not just important — but decisive.

For front-office professionals, the differentiator in an AI world will not be computational skill. It will be presence, empathy, connection, and trust. As I often tell audiences, "AI cannot replace presence. AI cannot replace rapport. AI cannot replace relationships."

Those technological limitations become the new terrain of professional advantage. In a world of AI, relationships become an even more effective moat that protects your career and a multiplier that can accelerate it.

I saw a glimpse of this shift back in 2017, years before generative AI exploded into the mainstream. At the time, a junior analyst working for me was struggling with a research project. I had told him exactly where to find the data — on the Bureau of Labor Statistics website — and returned a week later expecting an update. Instead, he told me he still hadn't been able to locate it.

So I asked, "Did you call the analyst listed on the website?"

He looked at me with genuine surprise.

He didn't know that calling someone was an option. It simply hadn't occurred to him that a real person, with a real phone number, was available to help. Like many younger professionals, he viewed calling someone as intrusive, almost inappropriate. So, he spent an entire week searching for data he could have obtained in minutes.

After our conversation, he finally picked up the phone. He called. He got the information almost immediately. And then, with a sense of revelation, he told me he planned to share this "secret" with his friends — this secret that you can simply call people when you need something. Today, he is a successful tech executive, and I am so proud of him. I also suspect that he would tell you this was a pivotal learning moment in his career.

It's All About People

AI can search. It can summarize. It can analyze. But no AI model can replace the value of reaching out to a real person, forming a connection, or building a relationship. That's uniquely human.

I learned the importance of connection early in my own career. My first job was in the summer of 1996, working in the Disney College Program at Walt Disney World in Orlando. One lesson was repeated over and over again during training: it's all about people. It was true then. It is true now. And in the age of AI, it will become more true than ever before.

Unlimited analytics will not eliminate the need for human relationships — they will elevate it. The ease with which AI handles information means professionals should devote more time and energy to the one thing technology cannot do: connect.

A decade later, I would learn that lesson again — this time in a way that shaped the future of my business.

A Business-Saving Personal Touch

In 2009, in the depths of the worst recession since the Great Depression, I started Prestige Economics. By the fall of 2010, my $150,000 nest egg had dwindled to about $4,400. My business was on the brink. With some of the very last money I had, I made a choice that many would have considered irrational: I spent it on handwritten holiday cards.

Everyone else that year was sending digital cards — quick, convenient, impersonal, and forgettable. In contrast, I hand-addressed hundreds of envelopes and wrote sincere notes thanking people for our work together, letting them know I appreciated them, and sharing thoughtful ideas about how we might collaborate in the future.

Those cards were not automated.
They were not optimized.
They were not efficient.

But they were human.

In January 2011, one of the CEOs who received a card hired me for a major consulting project. That single engagement saved my business. And from that experience, I took away an enduring truth: in a world where everything can be automated, relationships are the last frontier of competitive advantage.

Every year since, I have sent hundreds of handwritten holiday cards to honor the memory of that moment that saved my business.

When my books have been released, I have done the same. And when *The Future After AI* is released, I will send hundreds of copies of the book to people who matter to me and to my work. Taking time to let people know you're thinking of them is more vital now than at any point in modern professional life, and yet still nowhere near as important as it will be in the years to come.

Strategic Recommendation: Plan to Connect More Directly

Generative AI will transform operations, automate analysis, and organize workflows that once required teams of people. But paradoxically, this revolution in automation will force the front office to become more human, not less.

Trust, presence, empathy, listening, follow-through, and authentic connection will define the next era of professional success. AI may power the analytics, but humans will power the relationships. And the relationships will power the business.

As you look at the risks and opportunities of AI, I would implore you to identify three ways to connect directly, personally, and authentically with your clients in the year ahead.

And keep in mind —
Who you know doesn't matter nearly as much as who knows you.

CHAPTER 10

TARGET YOUR HIGHEST BILLABLE HOURLY RATE

The GDP and productivity implications of AI for the macroeconomy and the labor force are huge. But for individual workers, the challenge is different.

Each of us will need to move toward our highest-value billable hourly rate. In other words, what is the thing you do that adds the most value to your work? If you were literally paid by the hour, what part of your job would command the highest billable rate?

Once you know the answer, the next question becomes how to make that work a bigger part of your job. And how can you digitally delegate the lower-value tasks — anything that can be done by almost anyone — to AI?

This is the core shift.

Success depends on engaging in increasingly sophisticated digital delegation so your time is spent on the most advanced, highest-value work that only you can do.

Moving Up the Skill Curve

To protect your career and maximize your economic value, you need to continuously move yourself up the skill curve toward the work that cannot be easily automated, replicated, or delegated. This means focusing on the tasks that require judgment, experience, trust, and human connection.

An important example of where the highest value a profession delivers can be seen is in the world of estate attorneys in the run-up to 2010. In 2010, the U.S. federal estate tax — otherwise known as the death tax — went away for one year only. It gradually declined until it hit zero in 2010, then returned in 2011. But at the time, there were discussions that the elimination might become permanent.

At the time, my father and I had lunch with an estate planning attorney he had worked with for years. My dad asked, "What would you do if the federal estate tax goes to zero?"

The attorney looked at him and said, "Jeff, don't worry. There will always be prodigal children, philandering spouses, and unreliable business partners. I will have a job forever."

He was not joking. His point was clear. The highest-value work was not filling out estate tax forms. It was managing people, relationships, and conflict. It was work that only he could do.

The Real Value Lawyers Provide

I was reminded of this recently after giving a presentation to the executives of a top-tier law firm that delved into deep strategic topics about the global economy, AI, and geopolitics.

It's surprising to me how many people mistakenly think that corporate lawyers just add value by writing contracts.

After all, contract drafting can increasingly be done with AI, since contract drafts and templates are abundant and have existed for as long as computers have. They are infinitely searchable and easily draftable because contract law is based on hundreds of years of case law.

Discovery is also not the greatest value-add for lawyers, and there have been companies working to do this with AI for years.

That's why the greatest value lawyers bring is much more sophisticated and personal than contracting and discovery. Lawyers add the most value when they act as a buffer between counterparties who might otherwise end up in conflict, when they hold people's hands through difficult decisions and negotiations, and when they protect clients from themselves.

Yes, AI will draft more agreements. But I would still encourage anyone to use an attorney, because their highest-value role is the human one: creating good boundaries. To invoke the line from the Robert Frost poem *Mending Wall*, "good fences make good neighbors." Lawyers add a lot of value as fences and walls.

In corporate law areas like M&A or A&D, the highest-value skills are not document preparation and review but identifying strategic pitfalls, anticipating risks, and guiding executives through complex choices. In governance and compliance, managing and limiting systemic and reputational AI and cyber risk governance will be increasingly critical.

Being a good lawyer in the world of AI will be increasingly about managing relationships, navigating emotions, reducing risk, and providing strategic guidance. These will be some of the same highest-value skills and tasks for leaders across industries in a world of AI.

The Universal Need to Move Toward Highest-Value Work

Across every profession, focusing on the highest-value work will reveal where the top opportunities and priorities lie in a world of AI. Identifying the part of work that creates the most value and embracing and elevating that portion of your role will be the key to success.

With an unlimited bullpen of analytics at your disposal, analysis that once required specialized knowledge can increasingly be done with the aid of generative AI. There are key areas that cannot be replaced by AI, and they are intrinsically tied to how we relate to each other as people.

Relationships cannot be replaced by AI, trust cannot be replaced by AI, and judgment cannot be replaced by AI.

Across most industries, the highest-value work is often in the relationships you maintain with clients, with vendors, with customers, with employees, and with shareholders. The relationships are where the irreplaceable value lives. Everything else, everything that is not your highest-value contribution, is the part most likely to be automated, optimized, or delegated through AI.

Even for strategic planning, which you may consider a core competency that should not be digitally delegated, generative AI can be helpful. After all, generative AI can draw on a repository of historical examples, but it can also help pressure-test your own assumptions.

The Path Forward

The future of work will reward those who embrace digital delegation while also intentionally moving to advance the tasks only they can do.

It will reward individuals who clarify their highest billable hourly rate, expand that part of their role, and aggressively delegate everything else. And it will reward the value of the relationships you build and foster. More than ever, your network will be your net worth.

This is how you protect your career and thrive in the era of AI.

Strategic Recommendation: Reallocate Your Time

My top strategic recommendation for everyone reading this book is to follow three steps to spend more time on your highest-billable-hour activities and take your career to the next level.

First, identify the top three most value-add activities you do professionally. Chances are, those things involve helping other people and building relationships or they may involve technical expertise you have spent significant time developing.

Second, use AI to free up 10 percent of your time by spending less time on your least value-add tasks.

Third, dedicate that 10 percent time savings to spending more time on your highest value-add activities.

By focusing on high-value-add activities, you will be building skills that can help you distinguish yourself, increase your personal satisfaction, and, hopefully, make you AI-proof.

CHAPTER 11

BRING UNIQUENESS

To stand out in a world of AI-generated content, people need to focus on finding ways to truly bring uniqueness to their work. Anything formulaic, anything that fits a well-worn structure, is something generative AI can recreate with relative ease.

Shortly after ChatGPT was introduced, I remember showing it to a friend, who is a screenwriter and movie director. At that point, he hadn't yet heard of it. It was late 2022, just after its release. To demonstrate the potential of generative AI, I prompted ChatGPT to come up with a movie title for a story about a werewolf and a vampire in love in Texas, and it immediately generated the title *Forbidden Blood Moon Over Paris, Texas.* Then I had it write a few scenes.

The scenes were extremely generic, nothing out of the ordinary, and I could immediately see why the film industry would worry about this. Of course, a star-crossed love story between a vampire and a werewolf is a mix of deeply mined tropes. With an enormous existing repository of those kinds of stories, the model simply recombined pieces of what already existed.

Creativity in the Rear-View Mirror

Economists and statisticians often describe relying on historical data to predict the future as trying to drive down the road while looking only in the rear-view mirror. That same analogy applies to creative work as well. If what you create is generic and not innovative, AI can duplicate it. And this reality is already reshaping the digital landscape.

As of November 2024, for the first time in history, AI generated more than half of all new online content, according to Graphite.[1] That share has been rising sharply, and the trend is accelerating. This has profound implications for how thought leaders, business professionals, and companies stand out.

When AI produces a growing majority of content, the relative scarcity and value of anything genuinely unique increases.

EXHIBIT 11-1[2]

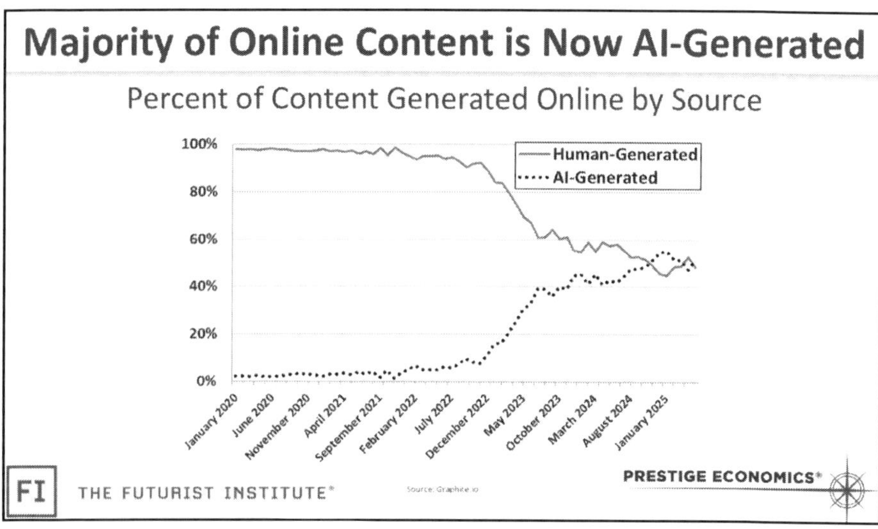

The Power of a Single Big Idea

The truth is that most business books have one core idea, one kernel that defines them. You can summarize many of the most influential business books in a single sentence.

- *The 4-Hour Workweek* is fundamentally about delegating tasks through online resources to achieve the lifestyle you want.

- *Blue Ocean Strategies* is about finding products or markets without competitors.

- *Getting to Yes* distills down the value of collaborative negotiation and structured communication.

- *How to Win Friends and Influence People* is about showing genuine interest in others so that affinity naturally increases.

- *Soar With Your Strengths* argues that people should focus on what they are good at rather than devoting excessive energy to eliminating weaknesses.

Each of these books is remembered for one core idea. And my own books follow the same pattern.

- *Cold War Two* rests on the argument that Cold War One and Cold War Two are two parts of the same conflict to prevent Chinese hegemony over Asia.

- *Jobs for Robots* centers on the growing prospects of automation that will fundamentally change the future of work.

Scarcity in an Age of Abundance

Every book comes down to one kernel of insight. In a world where AI generates the majority of global content, those kernels become even more important.

There will be fewer and fewer unique ideas relative to the overwhelming volume of AI-generated material, which will mostly mix, remix, and regenerate what came before. That scarcity increases the value of originality. For professionals, thought leaders, and businesses, having something unique to say — some original insight, perspective, or story — will become not just advantageous but essential.

The Timeless Value of Storytelling

And storytelling itself will matter more, not less. Even today, the biggest musical artists like Taylor Swift effectively give away their music for free, hoping fans will pay to see them perform live. Thought leaders write books that sell at low price points, hoping companies will hire them to speak.

But what is a speaker, fundamentally? A storyteller. Someone who can spin a yarn in a compelling way.

That ability to craft narrative, to hold attention, and to deliver a unique and memorable experience is going to become more valuable in the future because it is one of the few skills that remains difficult to replicate at scale.

Standing Out in the AI Era

If you want to be outstanding, you have to stand out. This is the core concept behind my LinkedIn Learning course *Building Your Visibility as a Leader*. The demonstrable value of standing out in your business, field, or profession has been proven in the past, and it will be even truer in the future.

The uncomfortable truth is that, in a world of AI, there will be much more content online, and that could make it tougher to stand out. But the encouraging truth is that there will be far less original content.

That means wordsmithing, idea generation, innovative thinking, and authentic storytelling will command a premium.

Strategic Recommendation: Share Your Stories

In a world of unlimited AI content generation, you need to invest time in developing your own unique frameworks, analysis, and perspectives. Increase your active original content generation to share new ideas you have created. Then, share your stories wherever possible — in LinkedIn posts, in online articles, or even in a book like this.

In a future where AI speaks much but says little, a person who can spin a yarn, speak in meaningfully new metaphors and analogies, and add uniqueness, texture, and originality, will be more inspiring and impactful than ever. Go out there. Deliver that impact and be that inspiration!

SECTION 3

IMPACTS AFTER AI

CHAPTER 12

THE FUTURE OF FINANCE AFTER AI

Artificial intelligence is transforming the global financial system, reshaping how capital is raised, allocated, audited, reported, and protected. In my LinkedIn Learning course, *The Future of AI for Finance and Accounting*, I talk about the high-value use cases for AI in finance and accounting that are already changing front-office, mid-office, and back-office work. The same forces are now redefining the broader future of finance after AI, creating enormous opportunities and serious new risks.

Finance at the Forefront of Technology

Finance has always been chasing and embracing a technological edge because that is where the money is. In the 1980s, firms that moved from paper ledgers to Lotus 1-2-3 and then Excel were not chasing a fad. They were fighting to boost productivity, speed, and profits. In some cases, they were fighting to survive. Today, the leap from spreadsheets and traditional statistical software to AI tools is just as crucial. Firms that use AI will gain speed, accuracy, scale, and a massive productivity lever in analysis and decision-making. Firms that eschew AI opportunities will lose market share and eventually fail or be acquired.

AI is more complex than a spreadsheet, but as I've laid out in other parts of this book, the logic for adopting AI is similar. A technology that can compile, clean, and analyze data faster than competitors becomes a structural advantage. Just as early adopters of electronic trading, risk analytics, and econometrics changed the competitive landscape, early adopters of AI in finance and accounting are already pulling ahead.

Software, Hardware, Integrated, and Generative AI

Most of the value that finance teams get from AI today comes from software. Statistical tools and machine learning systems search for correlations, trends, and outliers in financial and operational data. Programs like Stata, SAS, Power BI, and other AI-enabled tools democratize forecasting, scenario analysis, and profitability modeling that previously required teams with specialized quantitative skills and knowledge.

Hardware-based AI has also quietly been entering the world of finance. For years, drones and automated vehicles have been supporting physical inventory audits, facility inspections, and asset condition verification at remote sites.

Integrated AI systems connect hardware and software so that robots, sensors, and smart devices can gather data and immediately feed it into financial and risk models. Smart thermostats that adjust building loads, inventory bots that scan shelves, and cameras that count cars in a parking lot all create streams of information that finance teams can use to measure performance and refine assumptions.

Generative AI adds yet another layer.

Large language models can draft policies, procedures, contracts, and reports, supporting mid- and back-office functions. Generative AI can assemble first drafts of financial memos, variance explanations, audit workpapers, and investment hypotheses in seconds. Human experts still need to review and approve, while the hours spent on repetitive writing fall sharply.

Predictive Analytics, Forecasts, and Risk Management

Predictive analytics and forecasting sit near the top of the AI value pyramid for finance. Historically, only the largest funds and banks had the technology and talent to run high-end forecasting models or algorithmic trading strategies. Today, AI tools make advanced forecasting available to businesses of almost any size.

AI can ingest sales, pricing, macroeconomic, and behavioral data to forecast demand, margins, cash flows, and credit losses. It can track competitor prices or vendor quotes in real time, then suggest or implement dynamic pricing strategies. Models can be tested and retested continuously, so forecasts evolve as new data arrive rather than being updated infrequently.

Risk management may benefit even more. Machine learning systems can scan vast transaction histories to detect anomalies, spot emerging credit risks, and evaluate hedging strategies under thousands of simulated scenarios. In anti-money laundering (AML) and know-your-customer (KYC) programs, AI can flag structured transactions, smurfing patterns, and unusual fund flows far faster than human analysts. When the U.S. Treasury and large banks use AI to detect check fraud or suspicious wires transfers, they are turning pattern recognition into a real-time control function that protects customers and institutions.

Trading, Portfolio Management, and Investment Decisions

AI has already transformed portfolio management and trading. Complex quantitative strategies that were once exotic are now mainstream. AI models can build and rebalance portfolios around risk tolerance, time horizon, and factor exposures, while monitoring market conditions second by second. In high-frequency trading, algorithms have been executing orders for some time at speeds no human can match to capture miniscule price discrepancies and arbitrage opportunities.

Robo-advisors apply similar logic for retail investors, automating asset allocation and rebalancing at low cost. At the institutional level, AI can read earnings transcripts, analyze sentiment on social media, and process alternative data streams, such as satellite imagery and shipping records. All of this feeds into trade ideas and capital allocation decisions, and they will become increasingly important in the future.

Of course, there is a danger that widespread reliance on very similar models can create crowded trades and correlated behavior. When many funds use comparable AI systems, they may buy and sell the same assets at the same time. Over time, this could amplify volatility, stress liquidity, and generate flash crashes that reverberate through the financial system.

Data, Calculations, and the Hidden Work of Data Cleaning

Finance and accounting are fundamentally about data. Before AI can generate insights, that data has to be collected, cleaned, and structured. This data prep used to be one of the most time-consuming parts of the job — and it usually still is.

AI can help by scanning datasets for inconsistencies, missing values, currency mix-ups, or misaligned columns. It can flag anomalies quickly. Of course, human professionals should still be involved in the process to decide whether a value is an error, an outlier, or a signal.

Once data is in good shape, AI can automate repetitive calculations and support scenario testing. Whether the task is net present value, weighted average cost of capital, internal rate of return, or complex capital structure modeling, AI can run the math faster, under more scenarios, and with less manual effort. That frees people to focus on interpretation, judgment, and strategy.

The Future of Audit After AI

Audit is one of the areas where AI can be most transformative. Independent audits depend on testing and sampling large datasets to assess whether financial statements fairly present a company's condition. AI can review documents, reconcile transactions, and search for irregularities across every line item rather than only a small sample. In effect, the testing rate can approach 100 percent coverage in many areas.

Continuous monitoring becomes possible when AI watches transactions in real time, flags unusual activity based on risk scores, and routes exceptions to human reviewers. AI can prioritize high-risk accounts and locations, design risk-based sampling plans, and prepare much of the documentation required to support audit conclusions. It can also track evolving standards such as GAAP, IFRS, FASB guidance, and tax rules, reducing the risk of missing a regulatory change.

At the same time that AI offers opportunities, there is a growing risk that bad actors could use AI to become better at hiding prohibited transactions or engaging in other forms of fraud.

As detection tools become more powerful, sophisticated fraudsters and money launderers will adapt. They can use AI to break large payments into thousands of seemingly unrelated microtransactions, to route funds through webs of shell companies, or to exploit payment platforms and jurisdictions with weak supervision. Synthetic identities and deepfake documents can be generated at scale to pass automated checks.

This creates a paradox.

As audit testing and transaction monitoring push toward full coverage, the fraction of activity that is deliberately engineered to look normal may also rise. AI enables auditors to see more, but it also gives skilled adversaries the ability to simulate compliant behavior with greater precision. The result is a dynamic game of cat-and-mouse.

Success will require combining AI surveillance with human skepticism, robust data, strong controls over model governance, and close, ongoing collaboration with regulators and law enforcement.

Ethics and Governance Responsibilities

Ethics have always been central in finance and accounting because professionals in these fields are trusted stewards of other people's money. AI does not change that duty; it intensifies it.

Systems trained on historical data can reproduce or amplify past biases in lending, underwriting, hiring, or fraud detection. They can also become opaque black boxes that make recommendations no one fully understands.

Finance and accounting leaders need to demand transparency, explainability, and clear accountability for any AI system that influences financial decisions or reporting. They need to protect sensitive data with strong privacy and cybersecurity controls. They will have to ensure compliance with regulations on fair lending, market integrity, anti-discrimination, anti-money laundering, and consumer protection. Above all, they must remember that responsibility cannot be delegated to a model. AI is a tool; fiduciary duty and professional judgment remain human obligations.

Geopolitics, Macroeconomic Risks, and Economic Power

AI is being deployed in a world that is moving away from hyper-global integration toward a more fractured and contested order. Trade restrictions, technology controls, and investment screening are reshaping flows of goods, data, and capital. The decoupling between the United States and China is generating new patterns of foreign direct investment and supply chain design, while also creating new risks for markets and institutions that sit between competing blocs.

In this environment, AI becomes part of a broader economic security toolkit. It supports scenario analysis for tariffs and sanctions, improves stress testing for cross-border exposures, and helps firms understand knock-on effects when supply chains, commodity flows, or interest rate paths shift unexpectedly.

Countries that combine advanced AI capabilities with deep capital markets and robust legal systems will have a powerful advantage in the next phase of global competition.

Quantum Computing and the Limits of AI for Assurance

Looking ahead, quantum computing represents a potential step change in computational power, similar to the move from fire to nuclear energy. Quantum systems rely on qubits, superposition, and probabilistic, nondeterministic operations. That makes them extremely powerful for specific optimization and decryption problems, but it also means their internal workings do not always map neatly to traditional step-by-step deterministic logic.

For sophisticated traders and large institutions, access to quantum-supported analytics could improve trading strategies, risk calculations, and scenario modeling.

Yet the probabilistic nature of quantum computation may make it difficult to generate the kind of clear, reproducible workpapers that independent auditors, regulators, and courts require, especially when retail investors or public company reporting are involved. In many cases, any quantum-powered engine will need to be wrapped in additional systems that log inputs, assumptions, and outputs in a traceable way so that its use can be audited.

From Digital Delegation to Autonomous Agency

As of the mid-2020s, most AI in finance and accounting functions as digital delegation. Systems help people screen transactions, produce reports, draft documents, and run scenarios, but humans make the final calls.

Over time, the pressure to allow AI to operate with more autonomy will grow. An AI that not only analyzes ratios but also decides which metrics matter most and then executes trades, reallocates budgets, or releases payments without human review would be the logical next step in pure efficiency terms.

That is where the risks become greatest. Performance errors would scale quickly and could cascade across portfolios, markets, and corporate reporting. When AI controls action, not just analysis, the legal, ethical, and reputational stakes for boards, executives, and professionals rise sharply.

For that reason, even as AI and, eventually, quantum-supported computation become more powerful, the core responsibilities of finance and accounting will remain the same. Due care, due diligence, clear records, explainable reasoning, and strong controls will continue to define good practice.

The institutions that thrive in the future of finance after AI will be those that combine advanced technology with disciplined governance, a deep understanding of risk, and an unwavering commitment to integrity.

CHAPTER 13

THE FUTURE OF TECHNOLOGY AFTER AI

The future of technology after AI is one in which artificial intelligence becomes so prevalent that it fades into the background. AI will be a foundational element of the global technological operating system. It will be the assumed layer beneath software, hardware, services, devices, networks, and platforms across every industry. This transition is already underway — and accelerating.

A Long Arc of Data-Driven Innovation

When earlier innovations emerged, such as cloud computing, predictive analytics, machine learning, or early database systems, each advance marked another step in the long continuum of using data to generate value. AI represents the next step in that evolution. In earlier eras, companies focused on transitioning to the cloud or to mobile apps.

Today, AI is the buzzword *du jour*. But in the future, the mention of AI may decline, as it becomes so deeply embedded and universally deployed that it becomes unremarkable.

Value Extraction from Data

For decades, technology adoption has been shaped by one central theme: data. Organizations have been collecting data, using data, analyzing data, and extracting insights from data. AI in all its forms depends on the same effort. Whether it is generative models, agentic systems, or predictive analytics, AI takes data and turns it into value. This trajectory is not new. What has changed is that the scale and speed with which AI extracts value are dramatically higher than anything that came before it.

When AI Stops Being a Selling Point

Many startup founders today label their products as AI even when the technology behind them is no more than basic analytics. Although the label may be misused, it does not diminish the truth that the future of software is inherently data-driven. Nearly all future software will incorporate components related to data analytics, predictive modeling, or autonomous decision support. Whether people continue calling these capabilities AI or not will matter less over time. What will matter is the output and the value these solutions provide.

From Buzzword to Assumed Infrastructure

In the future, the phrase AI-enabled may also disappear in the same way that companies no longer advertise that they have a website. At one time, promoting a website mattered. Today, it is assumed. In earlier eras, companies promoted search engine optimization or cloud enablement. Over time, these technologies faded from public messaging because they were understood to be universal. AI will eventually reach that same point. It will be embedded and expected in daily digital life.

Quantum Computing as the Next Accelerator

Quantum computing will eventually scale to support and integrate with AI, unlocking new solutions for complex math problems and material science challenges. Most people will not understand quantum-enabled breakthroughs, but they will not need to understand quantum. In the same way that most people do not know how microprocessors work or how semiconductors are manufactured, quantum will operate beneath the surface. Only specialized engineers and scientists will work directly with these systems, even as society benefits from their advances.

Technology People Use Without Understanding

The same dynamic will apply to AI. People do not need to know how models are trained or how inference functions work. They will only know that the technology helps them. AI will allow workers to increase their output, improve the quality of their deliverables, and expand their capabilities across many domains. This is the fundamental value proposition. AI drives productivity, enhances decision-making, and generates new forms of economic value.

Cybersecurity and Endless Multifactor Authentication

With more data collection, ubiquitous AI, state-sponsored cyber threats, and emerging quantum computing decryption technology, there will be increasingly complex global AI imperatives and legal requirements (like GDPR) to protect personally identifying information (PII), financial information, work-in-progress intellectual property (WIP IP), any original digital assets of value, and access to online platforms. While there may be elegant solutions in the future, the most likely near-term solution is a brute-force increase in authentication requirements.

In recent years, there have been security initiatives to advance two-factor authentication (2FA), in which people use a combination of a password with a second factor of authentication, such as a passcode sent via a call, text, or email. Recently, I have already been asked for three-factor authentication (3FA) and even four-factor authentication (4FA), which required a password, a text code, an email code, and an authenticator app code. Unfortunately, given the rising cyber risks in a world of greater technology, the future is likely to involve multi-factor authentication (MFA) *ad absurdum*.

Invisible Tools That Shape Everyday Life

As AI becomes the operating system of countless industries, companies, and individuals will focus less on the underlying technical details and more on the results. No one pays much attention to their current computer operating systems. Nor do people care how combustion engines operate or how indoor plumbing works, but they depend on those systems every day. AI will become another one of these silent dependencies. It will be an essential part of daily life, understood by few yet invaluable to many.

A Future Defined by Quiet Ubiquity

Even if the public stops talking about AI, its economic value will continue to compound as AI becomes the foundational bedrock of the future economy. The future of technology after AI is a future defined by quiet ubiquity. AI will be everywhere, powering everything, without needing to be mentioned. Technology will fade deeper into the background, even as it creates greater value in the foreground. The use of the phrase "artificial intelligence" may diminish, but the impact of AI will continue to grow.

CHAPTER 14

THE FUTURE OF ENERGY AFTER AI

Artificial intelligence is transforming the global energy landscape. The rise of generative AI, agentic AI, automation, and data-intensive digital delegation has created what appears to be almost limitless demand for power. As companies rush to scale AI capabilities, global electricity demand is rising so quickly that traditional energy planning frameworks no longer entirely apply.

Domestically, U.S. energy demand in the near term is being shaped by the computational hunger of AI, along with a priority to reindustrialize the U.S. economy as a means of conflict deterrence to prevent World War Three. This shift is altering global markets, increasing the strategic importance of natural gas and nuclear power, raising concerns about sustainability goals, and accelerating geopolitical tensions in Cold War Two.

In the current and future environment, energy is poised to be a more crucial part of the foundation of economic growth and the fuel powering AI-enabled macroeconomic productivity gains, U.S. competitiveness, and even national security.

AI as a Global Energy Demand Multiplier

As I have noted throughout this book, generative AI has the potential to enable massive digital delegation at the population level across the global economy. However, every time AI takes on more work, more power is required to deliver those productivity gains. Even before the rollout of ChatGPT and other generative AI programs, not a single country had a high per capita GDP and low per capita energy consumption. In the future after AI, that is likely to be even truer.

Workers across the U.S. and global economies are effectively delegating tasks to algorithms, and those algorithms run on electricity. As a result, the marginal productivity of AI is tied directly to the marginal watt.

The implications of these dynamics are threefold.

1. The demand for power is poised to rise.
2. With an increase in demand for power, the price of power is poised to rise.
3. With increased power demand and costs, the cost of AI is poised to rise.

AI Power and Infrastructure Outlook

An August 2025 McKinsey analysis highlights the scale of these dynamics. Global demand for data center capacity may more than triple by 2030, with a compound annual growth rate of 18 to 27 percent and a mid-case of 22 percent. In the United States alone, demand is expected to grow 20 to 25 percent annually. These numbers imply massive ongoing energy requirements that go far beyond existing generation and transmission capacity.[1]

To support this build-out, companies are expected to invest almost seven trillion dollars in global data center infrastructure by 2030. According to McKinsey projections, more than four trillion dollars will go toward computing hardware, while the remainder will be invested in real estate and power infrastructure. Over 40 percent of all spending will occur within the United States, pushing American energy markets to their limits.[2]

McKinsey also notes that primary data center markets are already saturated. Vacancy rates fell to just 1.9 percent in 2024. The result is that hyperscalers are expanding aggressively into secondary and emerging markets, which often lack robust power and water supplies. As they scale, the pressure on public infrastructure is intensifying. U.S. data centers are expected to add about 460 terawatt-hours of electricity demand from 2023 to 2030, tripling current consumption levels. This will strain regional grids, challenge state decarbonization goals, and require rapid construction of new generation capacity.[3]

AI power demand is creating a structural shift in global power markets that will affect every energy source.

Natural Gas: The Quiet Winner of the AI Energy Boom

While renewables will grow, intermittency remains a defining limitation. After all, the sun does not always shine, and the wind does not always blow. AI needs constant power availability, not intermittent bursts. As a result, natural gas is emerging as the preferred transition fuel for the AI era.

Natural gas provides reliable baseload and peaking power with roughly half the CO2 emissions of coal. It is plentiful in the United States, flexible in its deployment, and geopolitically safer than relying on adversarial nations. Natural gas is also the fastest and most scalable option for meeting the enormous AI-related load growth expected this decade.

The rise of AI is likely to push natural gas prices higher and ensure strong demand for years to come. Nations that control natural gas will gain strategic leverage in Cold War Two. Those that do not will face rising energy costs, rolling blackouts, and weakened competitiveness.

Renewed Interest in Nuclear, Including Modular Reactors

AI demand growth is creating new momentum for nuclear power. After decades of slow development, nuclear is returning to the center of energy planning conversations in the United States, Canada, Europe, Japan, and South Korea.

Two forces are driving this shift.

First, AI-related power demand is growing continuously, and nuclear power provides a stable baseload without intermittency.

Second, AI demand is increasing the urgency of developing small modular reactors. Modular nuclear offers scalable, flexible, and potentially faster-to-deploy options for industries with intense power needs. Many companies are now exploring on-site or near-site modular nuclear reactors to power data centers and industrial loads.

The most AI-intensive companies will want control of their own energy supply, and modular nuclear could allow that. But the upfront CapEx and regulatory hurdles remain significant for greenfield nuclear facilities in the United States, including potential greenfield modular nuclear power reactors.

Behind-the-Meter and Off-Grid Power

The explosive growth in AI demand is driving many companies to invest in behind-the-meter power systems. These private generation assets, often off-grid or partially disconnected from regional transmission networks, could provide energy independence, ensure predictable pricing in volatile markets, and allow corporations to bypass grid constraints and permitting delays.

Behind-the-meter systems can include natural gas turbines, on-site solar or wind, batteries, microgrids, hydrogen fuel, and eventually modular nuclear.

Companies with heavy AI loads are already pursuing these strategies, and widespread adoption is expected. This means that the future of energy after AI will not just be about national or regional power grids. It will include private, corporate-controlled electricity ecosystems.

Sustainability Goals Under Pressure

AI is creating a paradox. On one hand, it promises massive productivity gains, including efficiency improvements in manufacturing, logistics, and research. On the other hand, the enormous energy demand required to train and operate AI models challenges emission-reduction goals.

Electric vehicles are central to many climate strategies, but they come with noteworthy vulnerabilities.

EVs rely on stable and inexpensive electricity. But as AI increases power demand and prices, EV cost competitiveness may erode. Higher power prices may slow EV adoption, especially when combined with supply chain risks for battery minerals.

EVs depend on lithium, cobalt, nickel, and rare earth minerals. China dominates the refining, processing, and manufacturing of many of these materials. Battery supply chains are deeply exposed to geopolitical risk, especially as Cold War Two intensifies. In the future after AI, the EV transition will be shaped as much by geopolitics and energy costs as by climate ambition.

There are other sustainability challenges presented by AI's rapid adoption and associated energy demands as well.

For example, in some regions, AI demand is already creating tension with net-zero targets. Some areas in the U.S. are also encountering lower reserve margins and anticipated power shortfalls. Plus, in drought-prone areas, water scarcity complicates the use of cooling systems for data centers and thermal power plants. Finally, there is also growing potential for more coal-fired power generation as a means to meet the almost insatiable power demands of AI.

These challenges present a growing risk that many companies may quietly reassess their sustainability goals if AI power demand pushes electricity prices materially higher.

Cold War Two and the Energy Security Imperative

The geopolitical landscape is increasingly defined by the alignment of China, Russia, Iran, and North Korea. Energy and mineral supply chains are central to their influence. Russia uses natural gas and oil as strategic weapons, while China controls renewable infrastructure supply chains and battery mineral refining. In Cold War Two, energy supply chains are weapons of national power.

Energy is a core component of geoeconomic strategy. It is also a prerequisite for national competitiveness in the AI era.

The Foundation of Prosperity in the Future After AI

There are no high-growth economies without significant energy consumption. Productivity gains from AI will be possible only if reliable and affordable power is available. The United States begins the AI era with a decisive advantage. It is more energy independent than during Cold War One, with deep natural gas and coal reserves, rising oil production, and expanding nuclear potential.

But the race is on.

Energy will be the foundation of economic strength in the future after AI. It is a strategic asset. It is a source of national power. It is the key that unlocks the next generation of growth and wealth. The nations that secure reliable energy supplies, develop resilient supply chains, and ensure stable electricity for AI will define the future.

CHAPTER 15

THE FUTURE OF HEALTHCARE AFTER AI

One of my earliest memories of visiting a doctor was going to an allergist in elementary school. His name was Dr. Bernard Berman, and he was a kind man who constantly used a Dictaphone. He dictated every note and every observation into a recorder that is the size of most desktop computer towers. After each visit, someone else had to transcribe his words by hand and enter them into a chart. It was slow, inefficient, and labor-intensive, yet it was the cutting-edge workflow of its time.

Today, all of that has changed. There is no transcription. There is no typing. Documentation is instantaneous. And in the future, it will only become more accurate, more seamless, and more deeply integrated into every corner of healthcare.

But even as AI transforms documentation, diagnostics, and administrative processes, the core of healthcare, which includes physical caregiving, will remain firmly human for a very long time to come. Healthcare will continue to be one of the fields where physically demanding, in-person roles dominate, despite the rise of artificial intelligence.

AI offers powerful productivity levers across mid- and back-office functions in healthcare, excelling at pattern recognition, analytical support, and diagnostic assistance. But many of the most urgent needs in healthcare going forward involve tasks that require human presence, empathy, judgment, and physical touch.

Elder care illustrates this clearly. Feeding, bathing, dressing, and supporting elderly or infirm patients demands a level of sensitivity and physical awareness far beyond what current or near-term AI systems can deliver. These interactions involve unpredictable movements, fragile bodies, and moments where a gentle human touch matters more than any technological breakthrough.

I often ask audiences a simple question:
Would you let a robot bathe your cat?

The answer is almost always no — unless someone *really doesn't like* their cat. Seriously, should not let a robot bathe your cat if you want to have a cat tomorrow.

The same logic applies even more profoundly to taking care of elderly patients.

AI can examine CT scans, mammograms, blood panels, and other diagnostic tests with extraordinary precision for pattern recognition, but expecting it to safely care for vulnerable patients is another matter entirely. And with populations aging across the United States and much of the world, the need for human caregivers is increasing rapidly.

Robots might one day fetch your dishes or vacuum your floor. But dishes aren't fragile in the way people are, and they aren't nearly as valuable. Your grandmother is both.

This demographic reality means that wages in healthcare will continue to rise. Forecasts reflect severe future shortages of doctors, nurses, home health aides, and other caregiving professionals in the coming decade. Even amid recent economic uncertainty and slowing U.S. payroll gains, healthcare employment has remained consistently strong. These jobs are essential regardless of what happens to GDP, tech valuations, or market cycles.

And as AI becomes ubiquitous, these workers will increasingly rely on tools that make their jobs easier. Dictation for nurses, hospice workers, home health aides, and social service professionals will reduce paperwork and free more time for patient care. Tasks that once required hours of writing or form-filling will be completed in seconds. AI will not replace these caregivers, but it will support them — allowing them to focus on the human elements of their role that cannot be automated.

Yet even with substantial AI-enabled efficiency gains, there still won't be enough people. The sector's labor shortage is nearly inelastic, and one of the few reliable ways to draw more workers into healthcare is to raise wages. As AI compresses entry-level roles in other industries, more workers may naturally shift toward healthcare because the pay premium becomes too large to ignore.

After spending more than two decades working in finance, I know one thing with absolute certainty: all markets eventually clear. Healthcare will be no exception. But it will take time, rising wages, and AI that reduces administrative burdens while preserving the essential human core of caregiving.

CHAPTER 16

THE FUTURE OF EDUCATION AFTER AI

Blue books are the future of education in a world of generative AI. Let me explain.

For most of the last century, American college and graduate students took exams using blue books — simple, stapled booklets in which students hand-wrote their answers, sometimes crafting complete essays. Even as universities adopted computers, computer labs, and eventually required students to own laptops, blue books remained the standard for testing. This relic of the past is likely to be the cornerstone of ensuring the future impact and efficacy of education after AI.

In 1995, the year I began college, my university required students to have computers for the first time. Before that, students used typewriters or shared computer labs for their written assignments.

But exams still happened in the classroom, on paper, under the watchful eyes of professors, TAs, or proctors.

For many years, classrooms once filled with students taking notes with pens and paper have become the domain of students using laptops. Despite this increased use of technology, we need to ensure students can learn and demonstrate their knowledge without relying on AI tools that write for them.

The Temptation and Challenge of AI-Assisted Work

Generative AI has made it incredibly tempting for students to use AI to write papers and complete assignments. This raises a fundamental testing issue. How do you test whether students can write without AI? The testing should occur in an environment where AI cannot intervene. That likely means more in-class, hand-written work — bringing us back to the blue book.

A professor friend of mine is an adjunct who faced this dilemma when ChatGPT first emerged. He had a choice: ban AI or allow AI. In the end, he chose to allow students to use AI — with consequences. He made it abundantly clear that the students were responsible for the quality and accuracy of the final work. In other words, use AI at your own peril. This approach can work at certain levels of higher education, but it still doesn't test whether students know how to learn, reason, write, or communicate effectively without AI assistance.

The Math Analogy: Calculators vs. Computation

At numerous points in this book, I have invoked technological changes in the financial world, especially the use of digital spreadsheets beginning in the 1980s. Just a decade before that, pocket electronic calculators became widely available in the 1970s.

Calculators and spreadsheets were tremendous productivity levers and quality checks for people working in finance, just as generative AI offers the opportunity to boost productivity and improve accuracy today. But technological innovations and opportunities sometimes present challenges for educators.

When thinking about the impact of calculators on mathematical education, I feel a strong desire to again invoke the popular economist mantra that *this time is never different*.

Having been born in the 1970s, I saw firsthand the temptation calculators presented to young students. They were generally banned during exams, and teachers would require most math tests to be taken without the aid of a calculator to ensure learners could demonstrate their capabilities without assistance.

Even today, although most students own a smartphone with a built-in calculator, tests for young learners still demand the demonstration of manual competency. The same future likely lies ahead for education and testing in the era of generative AI.

Assessments for analytical, content, writing, and communication capabilities seem likely to resemble how education approached mathematics: testing for competency without technological assistance. Homework might involve AI, but exams need to measure and test what the student can do alone.

The stakes are high because ubiquitous access to generative or agentic AI may lead to a decline in writing ability unless educators deliberately separate practice environments at home from testing environments in the classroom.

Advanced Education and the Role of Generative AI

At the higher levels of math, including college-level calculus, matrix algebra, or graduate-level quantitative work, calculators are usually permitted, expected, and sometimes even required. At those levels, students are solving conceptual problems, not practicing basic arithmetic. The same may eventually be true for writing, whereby graduate students are permitted or even expected to use generative AI as part of their research, revisions, and final outputs.

But this only works once students have already demonstrated that they can learn independently, reason effectively, and communicate coherently without assistance. Students still need to develop metacognition, structured thinking, and the ability to explain ideas in an unaided way. Those skills can only be properly tested in conditions where AI tools are not restricted and unavailable.

Moral Hazard of Generative AI

Generative AI can dramatically improve the polish of a report or the organization of an argument. But that improvement is not the same as demonstrating effective learning, analytical reasoning, knowledge, or communication skills. In fact, giving students access to AI too early may short-circuit the learning process entirely, just as a calculator for a young student could prevent them from actually learning multiplication tables.

The moral hazard is clear and significant: students may use AI to perform the work rather than to check their work.

This distinction creates a central challenge for the future of education: how to leverage AI to enhance learning without allowing AI to replace the foundational skills students need to develop.

A strong writer or researcher who uses AI becomes even stronger. But a weak or neophyte writer using AI risks outsourcing the core intellectual work entirely, producing outputs that reflect the machine rather than the student. AI can be a lever like calculators and spreadsheets to enhance human performance. But if it becomes a crutch, it undermines learning.

AI Throughout the Educational System

AI is already being integrated into education in many ways. It is being used for grading, performance analysis, scheduling, workflow, and administrative tasks. It is rapidly being incorporated into educational platforms and learning management systems. AI monitoring tools are also becoming common, raising their own set of debates and concerns. Ironically, if a student writes an essay report in a very structured and logical manner, it could be flagged by AI as AI-generated.

The Long-Term Risks for Learning and the Workforce

AI offers extraordinary opportunities to elevate writing quality and streamline communication. But if educators fail to preserve unassisted testing environments, the long-term result could be students who never develop core capabilities in the most basic skills, including reading, analytical reasoning, writing, and communication. These weaknesses could later manifest in the workforce, where the ability to learn new things is crucial for high-value contributors in an intellectual capital economy.

CHAPTER 17

THE FUTURE OF BUSINESS AFTER AI: B2AI AND B2A

For years, business leaders have focused on B2B and B2C commerce, referring to business-to-business and business-to-consumer sales. The next big shift in sales focus will be to B2AI and B2A commerce, which are business-to-artificial-intelligence and business-to-agent sales, respectively. Let me explain

Future B2AI (business-to-artificial-intelligence) commerce will force companies to shape their online presence, their content, and their marketing programs to target and gain the attention of AI systems. They will need to ensure that sales and marketing strategies enhance discoverability for generative AI programs.

As agentic AI becomes ubiquitous, AI-targeting strategies will become even more critical. AI agents will often act on behalf of individuals without including direct human input or intervention, which means business sales and marketing strategies will need to be more than just AI-discoverable. They will need to position offerings to sell directly to AI agents as B2A (business-to-agent) commerce, streamlining offerings to increase future sales directly between businesses and agents.

When the Customer Is Not Even in the Loop

Traditionally, a person searched for a product, booked a flight, compared prices, or looked for a gift. The human was always in the loop. But as agentic AI becomes more common, humans might be in the loop less often, or not at all. The AI might act independently because it has been granted permission to handle specific tasks or make certain purchases.

This raises a series of new questions for businesses.

How do you get discovered when the customer is not doing the searching?

How do you show up for an AI agent that is scanning options, comparing product data, or automatically looking for the best fit?

How do you get visibility when the decision-maker is a machine

At their core, B2AI commerce strategies are about how businesses attract AI systems and how business offerings are chosen by digital agents working on behalf of individuals.

SEO Becomes AIO

For years, people have invested in search engine optimization to improve discoverability online. SEO was about getting in front of human eyeballs. But the future is about getting in front of AI systems that may never show your content to a human if the AI does not think it matches the requirements.

This will require taking SEO to an entirely new level, where companies push toward AI optimization, or AIO. It's about making sure your business gets picked up, processed, and prioritized by generative AI systems.

In a world where B2AI and B2A sales and marketing channels get prioritized, it may no longer matter if offerings look good to people. Online content will need to be optimized so AI systems and their digital agents can interpret it, evaluate its relevance, and understand it at a structural level.

As agents take on more responsibility, this will become the foundation of product discovery. If your company is not visible to AI, it may be as if it does not exist at all.

A New Kind of Customer Experience

Throughout modern business history, companies have tried to build strong B2C relationships through outstanding customer service. They tried to build strong B2B relationships through professionalism and trust. Those still matter, but in the future, companies will also need a strategy for interacting with AI.

The AI is not a person. It is not emotional. It does not care about branding, tone, or personality. What the AI cares about is accuracy, clarity, structure, and alignment with the user's requirements. If the agent is looking for a product with specific features, it will scan thousands of options instantly. The business that gets the transaction will be the one that made itself easy for the AI to understand and easy for the AI to integrate with.

B2AI and B2A strategies will be about reducing friction for AI agents.

It will be about making transactions fast and clean, and offering customization in ways that digital agents can parse and evaluate. It is about presenting your offering in a way that makes sense to a system that might never read a website paragraph or watch a video. It is about meeting AI where it operates.

Agents as the New Market Power

As agentic AI becomes more mainstream, these systems will take on more tasks, including shopping for gifts, comparing travel options, buying groceries, or researching services. They will ask for very specific attributes on behalf of the users they represent. They will match product features to personal preferences, optimizing for time, convenience, and cost.

Companies that do not provide clear, precise product data will be ignored. Conversely, the companies that offer highly configurable options will stand out. And the companies that integrate tightly with AI systems — enabling transparent, swift interactions — will be the ones that win.

This transformation is not limited to retail or travel. It applies to almost every industry where AI can reduce complexity and eliminate friction. When the agent decides, the business with the best integration wins.

The Business Imperative of B2AI

B2AI is not optional. It is foundational for the future of commerce. It is the bridge between businesses and the AI agents that will increasingly manage the daily decisions of millions of people. It will shape how companies communicate, market, sell, and build relationships.

Businesses that act early will have the advantage. They will be the ones AI understands. They will be the ones AI chooses. They will be the ones that show up first in a marketplace where machines make the first move.

B2AI and B2A are the future of business. This is the next chapter of digital commerce. And it is the environment where companies must learn to speak not only to people, but also to the AI agents who now speak for them.

Even though B2AI and B2A sound like niche concepts today, they are early signs of where business is going. Automated agents will play a critical role in future markets. They will evaluate offerings, manage workflows, and execute transactions. This means companies will need to evolve in order to stay competitive.

The rise of B2A signals a major shift. In the future, your customer may not be a person at all. It may be an AI program, a digital agent, or an automated device acting independently. Businesses that understand this shift and prepare for it will be well-positioned in the years ahead.

CHAPTER 18

THE FUTURE OF CITIES AFTER AI

In the future, all cities will be smart cities. The pressures on urban leaders are poised to rise significantly in the years and decades ahead. Effective data collection and the use of AI offer some of the greatest hopes of easing those stressors. Most importantly, data can help improve efficiency and foster maximally perceived abundance, which will be crucial.

Rising Pressures

Urban leaders face increasing demands driven by higher rates of urbanization, greater caregiving needs, expanding tourism pressures, rapid technological change, and heightened geopolitical risks.

Demographic pressures have risen — and are poised to add more pressure in the future. The global population has doubled from around 4 billion in 1975 to 8.2 billion in 2025, and is forecasted to exceed 9.6 billion in 2050. Much of this population growth will occur in the least developed countries. More than half of the increase is expected to come from these regions, where some of the greatest increases in real GDP per capita are expected.

As population numbers rise and incomes grow, pressure on global resources will intensify. A larger and wealthier population will strain a Rousseauian world already constrained by limited resources. More people will demand and expect abundance in goods, services, and opportunities, and urban leaders will need to manage these pressures with strategic foresight.

Population and Urbanization Trends

Urbanization is accelerating. In 1975, the global urban population was about 1.5 billion or 38 percent of the world's total population. By 2025, it rose to 4.8 billion or 58 percent, and by 2050, it is expected to exceed 6.5 billion, which will be 68 percent of the global population.[1]

In less developed countries (LDCs), the rate of increase is even more dramatic. Urban populations in LDCs rose from 52 million in 1975 to 456 million in 2025 and are expected to exceed one billion by 2050.[2]

EXHIBIT 18-1[3]

Global Population and Urbanization to Rise

Total Population by Region	1975	2000	2025	2030	2040	2050	2025 to 2050
World	4,062	6,162	8,211	8,547	9,146	9,630	1,419
Least Development Countries	357	674	1,215	1,358	1,651	1,948	733
OECD Members	972	1,202	1,399	1,413	1,428	1,427	28
Europe & Central Asia	768	865	929	931	928	920	-9

Population and Urbanization	1975	2000	2025	2030	2040	2050	2025 to 2050
World Population	4,062	6,162	8,211	8,547	9,146	9,630	1,419
Urban Population	1,529	2,870	4,771	5,139	5,858	6,537	1,766
Urban % of Total	38%	47%	58%	60%	64%	68%	10%
Urban Population - LDCs	52	169	456	550	768	1,026	570
Urban % of Total - LDCs	15%	25%	38%	41%	47%	53%	15%

FI THE FUTURIST INSTITUTE® Source: World Bank PRESTIGE ECONOMICS®

This represents a rise from 15 percent of the population living in cities in 1975 to 53 percent by 2050. These kinds of rapid increases can create significant stress.[4]

New city dwellers, regardless of geography, bring new demands and encounter the pressures of densely populated environments. As people leave small towns and villages for cities, they increasingly look to local governments to provide solutions, support, and stability.

The Perceived Abundance Imperative

One of the most important responsibilities for urban leaders in the future is to use data to stretch the capacity of infrastructure, water, power, food, and supply chains. Leaders need to foster perceived abundance and, as much as possible, perceived unlimited abundance. When people feel constrained by shortages in rapidly growing cities, risks of urban unrest rise. Creating a sense of sufficiency and opportunity is central to long-term urban stability.

Demographic Shifts and Caregiving Pressures

Life expectancy has increased globally while fertility rates have declined. This means fewer young people and more older individuals who require greater levels of medical care and caregiving — and they will look to urban leaders for solutions.

Rising demand for healthcare services will affect every city. More jobs for nurses, doctors, caregivers, and associated roles will be essential. Ensuring that there are enough workers in healthcare is critical to maintaining perceived abundance and providing access to basic services and economic opportunities.

Remote Life as a Pressure Valve

Remote work, online education, digital commerce, and telehealth can relieve future urban pressures. These technologies allow people to benefit from urban services even if they do not live in urban environments.

Remote activities also support the perception of abundance by extending access beyond city limits, reducing stress on urban infrastructure and services.

Technology and Smart Cities After AI

AI will be crucial for fostering perceived abundance and maximizing the urban life experience of city dwellers in the future. To be effective, urban artificial intelligence solutions that drive cities to become smart cities will require good data — and a lot of it.

As in all other parts of life and society, AI requires good, big data. This means cities will need to deploy almost unimaginable numbers of sensors and engage in ongoing, unprecedented levels of data collection to stretch the value of infrastructure and assets to their maximum potential.

Strong fundamental data supporting the new AI operating system of urban environments will enable smart utilities, autonomous vehicles, drones, and automated systems that shape how cities operate in the future.

Of course, increased data collection need to operate in concert with resilient cybersecurity systems, making a solid tech stack conditio sine qua non for success.

Social Cohesion and Urban Stability

In addition to facing growing stresses on urban infrastructure and essential assets, urban leaders will also face mounting pressure to foster social harmony and cohesion.

Rising geopolitical tensions and broader risks create ever-greater threats to cohesion, especially with ubiquitous social media. Challenges related to misinformation, disinformation, false consensus bias, and subject identity will be among the most significant in the future after AI.

Real-World Best Practices from Seoul[5]

In my role as the AI and Emerging Technology Advisor to the Seoul International Policy Advisory Group (SIPAG), I have seen successful programs implemented by Seoul Mayor Oh Se-hoon as part of the Going Together with the Socially Neglected (GTSN) initiative. These programs are praiseworthy examples of how urban leaders and governments can provide support to city residents in a way that narrows social inequality gaps by focusing on education, income, and health.

For example, the Seoul Learn program provides free access to online courses and one-on-one mentoring for vulnerable students aged 6-24 who face barriers to private education due to socio-economic reasons. This initiative supports greater equality and fairness in education opportunities.

Among the 1,154 College Scholastic Ability Test takers who took Seoul Learn courses, 782 were admitted to universities. As of the end of July 2025, approximately 34,000 users were enrolled.

Seoul also has the Stepping Stone Income program, which helps fill financial gaps for low-wage earners (households earning less than 85% of the median income) to supplement the gap between their household earnings and the standard income level, boosting earnings without discouraging work in a way that allows income to grow for these households as they earn more.

Seoul also offers a program to support improved health outcomes. Wrist Doctor 9988 is a program that provides incentives for Seoul residents to be more active and support improved health outcomes through the use of smartwatches and a dedicated app for all Seoul citizens. This program, which is now Seoul's leading health initiative, has seen an increase in participation from 50,000 participants in November 2021 to 2 million participants by March 2025.

The GTSN initiative also includes Support for Socially Withdrawn Youth as well as a No More Loneliness Initiative to address issues of loneliness, isolation, and social withdrawal. These initiatives help to support social cohesion and urban harmony.

These examples from Seoul highlight how leading cities can prepare for a future in which every industry becomes a tech industry, and every city becomes a smart city.

To accelerate both general technological progress and smart city innovation, Seoul and the Republic of Korea have leveraged public-private partnerships to drive research and development, innovation, and investment.

Key Strategic Recommendations for Urban Leaders

The future presents major challenges — but also opportunities. Urban leaders can act now by focusing on the key strategies to foster perceived abundance, build strong cultures of data and education, embrace technology for smart city advantages, foster population health, and encourage social cohesion and stability.

Much will be asked of urban leaders in the future, but leveraging AI can help them deliver and improve the lives of their residents.

Facing pressures head-on will be essential for future success. For urban leaders, this means engaging in a sober assessment of a city's biggest challenges. Once those risks have been identified, urban leaders should determine what data can be collected related to those challenges. After all, problems that haven't been measured can't be improved.

Once sufficient data has been collected and analyzed, urban leaders should use it to intentionally inform policies with specific, targeted strategies that can also leverage ongoing data collection to measure and gauge efficacy. Over time, with a strong culture of data and data-informed policy initiatives, there is tremendous hope for progress to alleviate the most pressing urban challenges that have been targeted.

SECTION 4

STRATEGY FOR THE FUTURE AFTER AI

CHAPTER 19

THE FUTURE IS FOR FUTURISTS AFTER AI

Artificial intelligence presents significant potential for economic disruption, business uncertainty, and rapid change. In this high-stakes environment, futurists can provide more valuable insights and guidance than ever.

While many people react to change, futurists anticipate it, analyze it, and help others prepare for what's next. Whether working as analysts, consultants, strategists, or advisors, futurists bridge the gap between theory and practice by translating abstract technological, economic, demographic, and geopolitical shifts into actionable insights for businesses, policymakers, and leaders.

Navigating a world of expanding AI opportunities requires more than just reacting to events as they unfold. It demands structured scenarios that allow decision-makers to anticipate risks and opportunities, presented by economic and technological shifts. The ability to think critically about alternative futures and emerging threats is becoming an essential skill. It requires applying futurist frameworks to an uncertain situation. It requires being an applied futurist.

Thinking Like an Applied Futurist in the Future After AI

At its core, applied futurism requires asking what's likely to change, what will stay the same, and what major levers could shift the trajectory of AI use and adoption. Rather than predicting one single outcome, futurists create structured scenarios that allow decision-makers to navigate uncertainty.

Levers, Drivers, and Change Agents

Applied futurists examine the key forces shaping the future after AI, including proven ROI, relevant opportunities to gain productivity advantages, available computation power, and the future compounding technological advantages of automation, robotics, and quantum computing.

Every major transformation comes with risks and rewards. Futurists analyze the inflection points that could accelerate or disrupt industries, economies, national security, and global stability.

Fundamentals vs. Wild Cards

Some trends, including demographic shifts, economic cycles, and national security challenges from Cold War Two, are deeply rooted in history and are unlikely to change quickly. Others, like breakthroughs in quantum computing or black swan geopolitical events, could introduce rapid, unexpected change. The application of AI is not limited to a single economic sector. It is a technology with myriad value-add opportunities across a wide swathe of sectors in the U.S. and global economies, including finance, energy, education, healthcare, supply chains, and national security.

The role of applied futurists is to think across these dimensions and help decision-makers prepare for multiple potential futures.

Framing the Future

One of the most effective ways to approach strategic foresight is through the *Almost Now / Maybe Someday* framework. This distinction helps separate practical foresight for the near term from longer-term speculation, allowing leaders to engage with strategic planning without feeling overwhelmed by overly distant or uncertain possibilities.

Almost Now

These are the emerging trends and geopolitical realities likely to be widely adopted or intensified over the next decade. Decision-makers need to prepare for these shifts now, as they are actionable and immediate.

For corporate strategists, *almost now* technologies are ones that have already achieved minimum viable product and are likely being adopted by some major players.

Maybe Someday (or Maybe Never)

These are speculative but significant future developments that could reshape industries, global security, and economies beyond the next decade. While they may feel abstract today, ignoring them could be costly in the long run.

Another way to consider *maybe someday* would be as *maybe never*. In other words, this kind of scenario is unlikely to come to fruition any time soon. But it may never come to pass.

AI: Almost Now vs. Maybe Someday

For example, widespread adoption of AI to the point where it becomes the almost-invisible operating system and bedrock of the economy is going on right now, as an *almost now* phenomenon. However, a workforce without many people is a *maybe someday* — or maybe even a *maybe never* — scenario.

Cold War Two: Almost Now vs. Maybe Someday

I would also frame the transition to economic and technological decoupling between the United States and China as an *almost now* phenomenon, with a Technological Iron Curtain falling between the United States and China. The cleaving of dual use hardware and software technologies has been accelerating in recent years, and that separation between U.S. and Chinese technologies is poised to be almost total in the near term.

Despite an imminent potentially total split of global technological supply chains and resources, a full-scale bifurcation of the entire global economy, with entirely separate supply chains for *all* goods and services, remains a *maybe someday* scenario. And for some goods, like rare earth minerals or food, a complete division may be a *maybe never* scenario.

When talking about the future, there is often a propensity to lump technological, economic, and technological trend developments into one box called "the future." But knowing how to frame strategic discussions in a way that distinguishes these two time horizons ensures that organizations remain focused on relevant geopolitical imperatives while staying open to longer-term shifts.

Futurists as Thought and Strategy Leaders

As the speed and capabilities of technology accelerate and threaten to reshape the economy at increasingly rapid paces, the role of futurists will only grow in importance.

Leaders who engage with the future strategically to approach potential outcomes with high uncertainty will have a competitive advantage over those who remain reactionary. By applying structured foresight, organizations and governments can anticipate risks, identify emerging opportunities, and develop strategic resilience. Applying these skills and frameworks can add strategic value to an organization, help leadership priorities opportunities for a business, and shape the future of your own career.

The ability to analyze the relevance and potential impacts of technological, economic, and geopolitical trends, as well as consider alternative futures, and articulate strategic responses is will be increasingly valuable. When considering the potential for AI, these kinds of skills will become indispensable in a world of increasingly ubiquitous and powerful artificial intelligence capabilities.

The future is for those who prepare for it — and futurists will be the ones leading the way.

CHAPTER 20

FORECASTING, FORESIGHT, AND FOUR FUTURES AFTER AI

Forecasting has been the foundation of my career. Precision in predicting economic trends, financial markets, and business cycles has been critical to my work. But while forecasting is powerful, it works best in domains where historical data are abundant and reliable, data trends are clear, patterns are stable, and variables can be reliably modeled.

In contrast, some domains like revolutionary technology and geopolitical conflict are defined by complexity, uncertainty, and nonlinear developments. There are too many moving pieces to rely solely on traditional forecasting models.

This is where strategic foresight and futurist frameworks become essential. These tools help leaders, policymakers, and businesses prepare for multiple possible futures rather than being locked into a single deterministic outlook. The flexibility of mind and creativity required to create futurist scenarios are uniquely human skills that can be developed and improved.

One of the most effective frameworks for navigating uncertainty to craft futurist or strategic foresight scenarios is Jim Dator's Four Futures Framework and John Smart's adaptation thereof, which categorizes future possibilities into four key archetypes: continuation, plateau, reversal, and parabolic rise.[1]

Continuation of Trend

In this scenario, current trajectories persist, with existing economic, geopolitical, and technological dynamics unfolding in a relatively predictable manner. For AI, this could mean a continued adoption of AI and growing capabilities, accompanied by fewer errors and hallucinations, along with ever-greater value-add outputs. Businesses would adapt to the "new normal" of ongoing productivity gains accompanied by growing ROI.

If I had to name this scenario as a way to frame it, I might call it "More and better AI over time."

EXHIBIT 20-1[2]

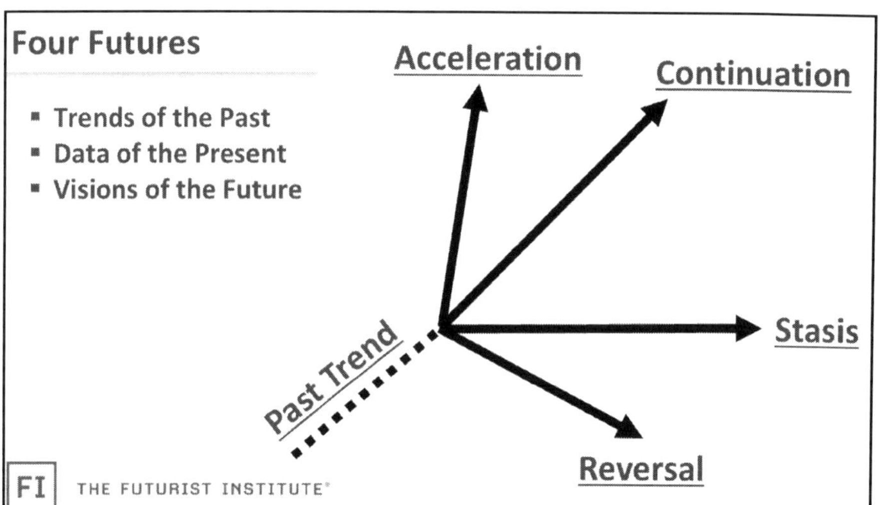

Discipline (Plateau)

This future envisions a world where external or internal constraints force a stabilization of trends. It could involve a scenario where global institutions, regulatory policies, domestic pressures, or resource limitations (for something like power or water) force the development of AI to freeze at its current level.

Throughout this book, I have often made the comparison of AI adoption in professional services to the value-add of digital spreadsheets. Spreadsheets have not changed fundamentally since the rollout of Lotus 1-2-3 or its follow-on replacement by Excel. Yet their use has become more widespread.

An AI Plateau scenario might see no further significant gains in AI capabilities. But even then, I would anticipate a growing use of AI that would still bolster massive economic and labor productivity gains in the years and decades ahead.

For businesses, this would mean an era of rolling out the existing technology to more workers with greater training, efficiency, and capabilities. However, more upside for future capabilities would be limited. In the world of finance, the full value of digital spreadsheets remains not fully captured. After all, consider how few people have mastered pivot tables and other moderate to advanced Excel functions.

If I had to name this scenario as a way to frame it, I might call it "We already have all the AI we will ever have."

Collapse (Sharp Reversal of Trend)
In this scenario, AI adoption could retreat. This would require an unusual set of conditions where the technological and labor market benefits of artificial intelligence become deprioritized or even stigmatized. The trend of technological adoption could face regulatory, economic, environmental, or other pressures to back away from — or push back against — AI.

This future would require business leaders to prepare to unwind their already-growing reliance on artificial intelligence and related technologies. If I had to name this scenario to frame it, I might call it "Backlash against AI."

Transformation (Parabolic Rise)
This future involves breakthroughs, major geopolitical shifts, or technological revolutions that dramatically alter the landscape.

In our AI scenario, this could mean achieving much more rapid or comprehensive AI capabilities, with much higher levels of fidelity and reliability. In this scenario, businesses and policymakers would need to be agile to capture opportunities in a rapidly evolving AI landscape that moves faster than we are currently anticipating. The output gains would be massive. But the labor market dislocations could also outpace anything we are currently pricing into our forecasts and policy scenarios. AI developers and adopters would be even bigger winners, but those who are slow to adopt these technologies would likely be much bigger losers.

If I had to name this scenario as a way to frame it, I might call it "From AI boom to AI sonic boom."

Weighing the Four Futures

Non-futurist consultants or analysts often frame future scenarios to include three cases: a high case, a mid case, and a low case. Futurist frameworks like the four futures I have just shared offer more opportunities for nuance — and the potential to identify key levers and drivers of outcomes. By recognizing the factors that will shape the future, it more fully prepares leaders to gain strategic advantages in the future and usher in what futurists call a "preferred future."

But there is an important caveat, even when it comes to strategic foresight and futurist scenarios: the range of scenarios may be wider, but there is still an anchoring bias to the current trend.

Although finance professionals most often invoke the phrase "the trend is your friend," it is easy to argue that most business leaders and strategists face a moral hazard that they, too, may succumb to the lure of assuming an ongoing trend will continue. This is what makes identifying the levers and drivers of scenarios infinitely more valuable than crafting the scenarios alone.

STEEP Framework

When considering the different ways that the future can be shaped, there are a few different ways futurists talk about this. One of the most common is the STEEP framework, which stands for Social, Technological, Economic, Environmental, and Political. For AI, the likelihood of a scenario where "more and better AI over time" seems quite high. But there is a wide range of factors — social, technological, economic, environmental, political, regulatory, and geopolitical — that could increase the probability of another scenario.

In general, the future after AI will be a time that is likely to be defined and dominated by elevated uncertainties of outcome that will be directly fomented by increased technological deployment and reliance.

Futurist Frameworks as Imperatives

The future after AI will likely present greater risks and opportunities than we face today, which underscores the imperatives of futurist frameworks.

For businesses, futurist frameworks allow leaders to reimagine growth opportunities, as well as risks and limitations, to prepare for success in multiple competitive landscapes.

For investors and financial markets, scenario-based planning enables better risk management, portfolio diversification, and long-term resilience.

Plus, strategic foresight can help policymakers prepare for shifts in economic sector activity, structural unemployment risks, growing Cold War Two cyberwarfare threats, supply chain vulnerabilities, and other national security risks.

Applying Strategic Foresight

While no one can predict the exact course of technological development and adoption, the ability to think across multiple futures ensures that leaders and decision-makers are not caught off guard. In a time of rising technological capabilities and more rapid change and uncertainty, those who embrace strategic foresight will not only survive but thrive.

CHAPTER 21

AI IMPLICATIONS FOR COLD WAR TWO

Proactively building connections with people is a critical imperative in the world of AI. But managing the risks of digital relationships and social media in the world of AI will become more challenging as it presents more significant risks.

AI presents significant positive opportunities for businesses and the economy. But it also presents growing downside risks in domestic and global culture wars playing out on social media. In recent years, social media has become one of the most potent tools for subversion, manipulation, and psychological warfare in Cold War Two — the conflict between the United States and its allies on one side, and China and its allies on the other.

Unlike traditional propaganda, which historically required significant resources for printing, broadcasting, or distributing physical materials, today's information warfare can be waged more cheaply than ever on social media. Plus, generative AI has already shown itself to be an effective tool that can be misused and deployed cheaply at scale to further drive tensions, sow seeds of discontent, and foster resentment and animosity.

Influence can be purchased at a cost per click measured in cents, and evildoers can leverage inexpensive online access and a basic understanding of engagement algorithms to spread discord and weaken opponents.

This economic structure incentivizes the creation of content that fragments society and fuels division. It takes advantage of the human tendency toward false consensus bias, convincing people that fringe beliefs are far more common than they truly are. The result is the widespread adoption of subjectivist realities that distort how individuals interpret the world around them, leading them to view political discourse only through an echo chamber.

The AI-Accelerated Narrative Machine

In my book *Revolt and Revolution*, I examined how compelling narratives have historically fueled revolutions by mobilizing the masses. The key to success is always the same: craft a message that taps into real or perceived oppression, economic hardship, or national humiliation, deliver it in a way that evokes strong emotional reactions such as anger, fear, resentment, or pride, and repeat it consistently across multiple channels to create the illusion of widespread consensus.

In Cold War Two, this process has been supercharged by social media algorithms designed not to inform but to provoke engagement at all costs.

Russia, China, and Iran have adopted similar tactics, utilizing AI-generated content, deepfakes, and bot networks to manipulate global discourse and disrupt Western cohesion.

Risks Growing Due to AI

What makes the current situation more dangerous is how dramatically AI can accelerate the creation and distribution of misinformation and disinformation, enabling adversaries to manufacture high volumes of persuasive falsehoods at scale. In the past, crafting realistic propaganda required expertise, time, and substantial manpower. With modern generative AI systems, an adversary can instantly produce credible fake images, coherent narratives, and fabricated evidence. These AI products do not simply amplify division. They can sow discord and generate divisions that did not previously exist. AI can create entire synthetic movements, imaginary controversies, and artificial public figures whose only purpose is to influence real people and alter real-world political and social outcomes.

The emergence of AI-generated misinformation represents a significant evolution in psychological warfare. Traditional propaganda focused on shifting opinion. AI-driven propaganda focuses on reshaping reality itself. When people are exposed to personalized falsehoods that match their individual fears, biases, and vulnerabilities, they begin to inhabit entirely different informational universes. Inside these personalized bubbles, contradictory or competing facts become nearly impossible to reconcile.

When AI enters this feedback loop, the effect becomes even more pronounced. AI systems can quickly identify which emotional triggers are most effective for each individual. They can then refine and personalize messages to maximize impact. This transforms digital agitation from a blunt instrument into a precision-guided psychological tool.

Cold War Two, AI, and the Rise of Digital Fifth Columns

The most dangerous consequence of social media weaponization in Cold War Two is the creation of internal divisions so deep that they function as digital fifth columns — communities that unknowingly act against their own national interests.

Unlike traditional warfare, where an external enemy is clearly defined, social media warfare creates self-sustaining cycles of distrust among citizens, erodes common national identity, and collapses the concept of objective truth, leaving societies fractured and vulnerable to foreign influence.

This phenomenon is easy to understand through the lens of Benedict Anderson's foundational work, *Imagined Communities*, in which he argued that nationalism was historically forged through shared language, print media, and collective identity-building efforts.[1]

In contrast, today's digital print capitalism creates ideological echo chambers shaped by algorithm-driven reinforcement loops that prioritize emotional provocation over factual discourse. Rather than uniting people under a common national identity, social media fosters hyper-segmented identities, some of which are actively promoted by foreign adversaries. With AI-enabled actors, these risks are poised to be even greater in the future.

Empathy Exhaustion and Creativity Erosion

One of the least discussed but most damaging aspects of social media weaponization is how it erodes both empathy and creativity, making societies more vulnerable to manipulation.

Empathy exhaustion is a consequence of the constant bombardment of crisis, outrage, and conflict. People become desensitized to suffering, making it easier for propaganda to succeed without triggering moral resistance. Instead of fostering deeper human understanding, social media conditions users to respond with quick, polarized emotional reactions that further entrench divisions.

AI also creates new problems by generating false realities at machine scale. Synthetic audiences and fake comment threads can give the impression that certain viewpoints are widely supported even when no real community exists. Deepfake videos can create the appearance of events that never happened. And AI-generated influencers can pose as authentic humans while quietly nudging opinions. Because these techniques can collectively foster a false sense of consensus, people come to believe that their views are validated and held by the majority.

Social media also further encourages emotional responses rather than thoughtful reflection. Instead of fostering understanding, it reinforces impulsive reactions that deepen social divides.

Creativity erosion is another unfortunate consequence of the modern digital environment. Social media platforms are designed to shorten attention spans and discourage deep analysis. Viral content tends to be simplistic and emotionally charged rather than complex and intellectually challenging. When AI systems optimize content for engagement, the effect can become even more extreme. People receive a continuous stream of emotionally provocative material that leaves little room for independent thought or creative problem-solving.

In Cold War Two, this erosion of creativity makes societies less adaptable and more vulnerable to manipulation. Adversaries benefit when populations respond emotionally rather than rationally, because emotional decisions are easier to manipulate.

Looking ahead, social media will remain a central battlefield in Cold War Two. It is one of the key areas where a Technological Iron Curtain is poised to fall more completely.

Foreign adversaries will continue to manipulate public opinion, destabilize democratic systems, and undermine trust through digital ecosystems. Governments have taken steps to mitigate these risks. But more needs to be done, especially as risks grow with the cheap ubiquity of artificial intelligence-enabled messaging.

The United States, for example, has banned TikTok for federal employees and introduced new regulations on AI-generated content. However, these measures do little to address the underlying problem. As long as platforms rely on engagement-driven business models, the incentive to promote divisive, emotionally charged content will remain.

Artificial intelligence ensures that misinformation will become faster to produce, more difficult to detect, and more psychologically impactful. The battlefield has shifted. The war is already underway. In Cold War Two, the fight for truth, national identity, and digital resilience will define the future of free societies. This is why increased scrutiny of artificial intelligence is likely, and the Technological Iron Curtain will fall more completely.

CHAPTER 22

AI IMPLICATIONS FOR BUSINESS LEADERS

In June 2025, I was in Hollywood for the premiere of a movie for which I served as an executive producer. While I was in town, I grabbed breakfast at Canter's Deli with a friend who is the CEO of a large subsidiary of a public company. Over coffee and eggs, our conversation turned — as it often does these days — to artificial intelligence.

He asked questions I've heard many times:
What can AI do for me?
How do I learn how to use it?
How do I start using it right now?

The First Lesson: Just Ask

Within a few seconds, I opened ChatGPT and demonstrated exactly how simple AI adoption can be. First, I showed him how to type prompts. Then I showed him the more intuitive path — using the audio function to speak naturally to the model.

His reaction was immediate. For him, this was eye-opening.

For many executives, new technologies feel daunting. And the older we get, the more daunting each new wave seems. I shudder to admit how long it took me to become a regular user of QR codes — and I work in future-facing fields. Longstanding habits and limited time make hesitation understandable.

But with generative AI, the barrier to entry is astonishingly low. The use cases are everywhere. The learning curve is gentle. And the value potential is enormous.

AI Isn't Complicated — And It Isn't Optional

This is not a technology that requires years of study to uncover its practical use cases or be able to use.

Generative AI communicates in natural language and responds intuitively. Yes, the tools differ in quality. I personally use the higher-end version of ChatGPT — around $200 a month — which I find far more capable than the lower-tier versions. Others prefer Anthropic's tools. But most of the value in the AI ecosystem comes from a handful of dominant products.

Just as Bitcoin became the flagship cryptocurrency despite the theoretical strength of other platforms, ChatGPT took an early lead. In the enterprise world, Google's Gemini and Microsoft's Copilot are emerging as likely long-term winners as well.

But the identity of the winners matters less than the underlying truth: business leaders should adopt this technology now. Hesitation is the modern equivalent of being a finance professional in the 1980s refusing to use digital spreadsheets.

The Greatest Productivity Lever Since the PC

Generative AI may offer the biggest productivity boost for service-based businesses in decades. When computers and email were introduced, many predicted paperless offices. The reality was the opposite: email created a flood of digital noise and more paper than ever.

Generative AI is different. It lets individuals and organizations focus with laser precision on specific deliverables. When paired with a trained expert or domain specialist, AI becomes a force multiplier. Productivity gains are substantial, measurable, and transformative.

In my own companies, AI has reshaped mid-office and back-office functions — tasks I once handled reluctantly or delegated to junior analysts. I haven't hired a junior analyst for this type of work in years. I may never hire another. AI now does it better, faster, and cheaper.

Is AI a Bubble? The Wrong Question

A debate is underway about whether AI is or is not a bubble. But this debate misses the larger reality.

The prices we see today are almost certainly artificially low. AI is following the classic drug-dealer model: make it cheap, subsidized by private equity and venture capital, let people get hooked, then raise prices once they cannot imagine working without it.

And they will not be able to imagine working without it. Just as no one in finance today would voluntarily abandon digital spreadsheets, AI will become indispensable. When something becomes essential, pricing power follows.

Inelastic Demand

With massive labor and productivity benefits, AI creates the potential for inelastic demand. In economics, inelastic demand means that people will pay almost any price to continue using a product because the perceived cost of switching is too high. Once someone locks into an AI program and it becomes central to their workflow, their demand becomes less sensitive to price changes. They simply do not want to leave. The relationship becomes sticky.

These three factors, hassle, time, and money, shape decision-making. Maybe you can reduce one of them by switching AI platforms, but you will probably have to pay more in at least one of the others. That tension creates inertia. It makes the current choice feel safer, cheaper, and easier to maintain. As AI becomes more deeply embedded in personal workflows and enterprise systems, that inertia increases.

These dynamics also push companies to move quickly. There is a strong incentive to complete a land grab early. If a company locks users in before they build habits elsewhere, those users are far less likely to switch later. It is always easier to retain an existing customer than to acquire a new one, and the difference in cost can be significant. Customer retention costs are typically much lower than customer acquisition costs.

This creates real pricing power for AI companies. Products that seem inexpensive today are likely to become more expensive over time. The deeper the integration, the more people rely on a platform, and the more knowledge that platform accumulates about the user, the higher the opportunity cost of switching becomes. That gives AI providers greater room to raise prices while still maintaining demand.

The economic and pricing potential for AI emerges from the combined forces of opportunity cost, switching costs, inelastic demand, and customer stickiness. Once people are locked in, they stay. As they stay, AI companies will gain long-term pricing power. And user loyalty is likely to be high and sticky.

Why AI Will Become More Expensive

Right now, generative AI is cheap. Tools are free, freemium, or bundled. But once adoption reaches a critical threshold and workflows, processes, and enterprise systems rely on it, organizations will willingly pay much more.

And what is the alternative? Hiring more people?

Labor markets have grown more expensive. Skills shortages exist across industries. Hiring, training, retaining, and replacing workers all come with increasing costs. AI does not.

The Drug Dealer Model

A lot of new technology operates on what is often called "the drug dealer model," which means the first hit is free — because they want to get you addicted.

Often, with new technologies, private equity or venture capital money subsidizes these activities to gain market share and create sticky customer relationships.

There are numerous examples in other technologies. Remember when Uber was first introduced, and it was very cheap — and much cheaper than taxis. In 2025, I have seen Uber be a lot more expensive than taxis in a number of markets, especially at times of peak demand.

This dynamic of rising labor costs, tightening labor markets, and the growing indispensability of AI helps explain the sky-high valuations we see in the AI sector. In short, people are getting hooked on the sauce!

Opportunity Costs

Opportunity cost is one of the most well-known economic concepts, and it is based on the notion that all choices have costs and that consumers will choose the item that presents the greatest benefits, even if they give up alternative benefits.

AI adoption is being fundamentally shaped by opportunity costs — and the costs of alternatives, including switching costs. This is also happening at the program level.

Like many technologies, generative AI and other AI programs are likely to be extremely "sticky" with users. Once people begin using AI agents, they quickly build habits, workflows, and knowledge structures around them. That means choosing an AI platform becomes an economic decision with real trade-offs.

When someone becomes accustomed to working with a specific AI agent, the switching costs can be extremely high, and the alternative could come with a meaningful loss. You might give up the accumulated knowledge, training, preferences, and workflows that your current AI already understands.

The alternative might offer a slightly lower subscription price, but the marginal cost of switching in terms of time, effort, hassle, and reduced efficiency can outweigh any financial savings.

High switching costs are common in the tech world, and often include the time required to set up a new system.

Stickiness and TCO

This stickiness appears across many familiar choices. People stay with a particular credit card, a bank, or a cell phone provider. They stay loyal to either PC or Mac. They stick with Excel instead of switching to Google Sheets, or the other way around. The reason is not always price. Switching takes time, effort, and hassle. You might save a little money on one dimension, but you will likely pay more in at least one of the others.

I have seen this play out in enterprises that have moved from Slack to Teams, thinking they would save money. But the labor hours, lost efficiency, and frustration involved in moving every communication channel, every data thread, and every historical record can be enormous. Even if the price tag is lower, the hidden costs are steep. This is why it's important to consider what procurement professionals call the total cost of ownership — or TCO — when considering making a switch.

Clear imperatives can be drawn from the fact that AI is cheaper now than it will be in the future, and investing in AI, automation, and robotics will be crucial to counter labor shortages and maintain competitiveness in reshored manufacturing industries.

Given the future opportunities and risks around AI, some of the most valuable advice I give executives is to apply the PAR framework I developed for use with clients of Prestige Economics and The Futurist Institute.

The PAR Framework

The PAR Framework is a simple way to think about how leaders, companies, and organizations can approach future risks, especially those tied to emerging technologies. PAR stands for Preparedness, Adaptability, and Resilience. The name itself is inspired by golf, where "par" represents the expected standard. In the same way, PAR describes the standard leaders should aim for when navigating a world of rapid change and uncertainty.

Preparedness

Preparedness is the foundation. It starts with using futurist frameworks and futurist scenarios to think creatively about what could happen. Preparedness means looking beyond the obvious and considering a wide range of outcomes. Leaders who are prepared understand that the future involves many variables, and they take the time to explore those possibilities. Preparedness allows organizations to start from a position of strength because they have already thought about the risks, opportunities, and surprises that might arise.

Adaptability

Adaptability is the ability to move quickly and respond to new information, new technologies, and new environments. Some people might call this agility or responsiveness, but adaptability is a better description. It means you can change as needed. You can adjust to shifts in markets, technologies, funding cycles, or customer demands. Adaptability is about staying flexible and being open to consciously evolving your strategy and approach as the environment changes.

Leaders and organizations that are adaptable do not freeze when confronted with change. They move. They learn. They retool. They adjust course without losing sight of long-term goals.

Resilience

Resilience is the ability to bounce back when things do not go well. It is the capacity to survive the next tech cycle, hype cycle, funding round, growth slowdown, or recession. Resilience determines whether an organization can continue to march on through pressure, shocks, and disruptions.

Every future involves uncertainty and stress points. Resilient leaders and organizations absorb these hits and continue forward with energy, clarity, and purpose.

Operating on PAR

When an organization is prepared, adaptable, and resilient, it is operating at PAR. It is keeping pace with best practices for navigating a world defined by greater complexity and greater complicatedness.

Complexity means there are many different factors influencing the future. Complicatedness means those factors can be difficult to pull apart, understand, and trace to their core drivers. This is why futurist thinking is essential.

Whether you are running a company, a division, a corporate board, or working in national security, the PAR Framework offers a practical guide. Leaders of the future need to meet challenges at a high level of preparedness, adaptability, and resiliency. This is how they stay aligned with emerging risks and opportunities — and how they keep moving forward when the world keeps changing.

Embrace Opportunities Without Hesitation

Your teams are likely already using AI. And before long, AI will be much more expensive than it is today. This is the cheapest AI will ever be, although the value it delivers is only going to grow.

Business leaders need to do three key things

- Learn to use AI.
- Extract value from AI.
- Drive productivity with AI.

And you should ideally seek to do those things before your competitors. There is no time to waste. Opportunities are here for the taking.

CHAPTER 23

AI IMPLICATIONS FOR CORPORATE BOARDS

In an AI-first and AI-invisible world, corporate boards will need to operate with continuous oversight of emerging risks related to governance and operations, including data compliance, cybersecurity, audit integrity, communications strategy, and corporate reputation. More than ever, independent directors will need to bring strategic foresight to every decision, because the speed and scale of AI adoption create new vulnerabilities that can threaten the stability, credibility, and continuity of the enterprise.

Future board leaders will need to demonstrate an understanding of AI systems and their risks, ensuring that companies meet rising expectations for transparency, trust, and responsible innovation. Boards will need to effectively anticipate threats and guide leadership to safeguard against economic, geopolitical, and technological disruptions that could emerge without warning.

Governance in the Era of Intelligent Automation

Traditional governance models assumed predictable risk categories and relatively stable operating environments. AI has erased those assumptions.

Automated decision systems, autonomous tools, and AI-driven workflows require continuous governance oversight and a board that is on the loop. Directors will increasingly need to confirm that organizations have established clear accountability structures for AI use, oversight mechanisms, and documented processes to evaluate the impact of algorithmic decisions on customers, employees, and partners. Board members will need to incorporate potential impacts in the creation and enforcement of governance frameworks that include data stewardship standards, model development controls, and executive-level accountability for the ethical deployment of AI technologies.

Compliance in a Rapidly Evolving Legal Landscape

Companies need to comply with an increasingly complex web of regulations governing data privacy, cross-border data flows, algorithmic transparency, AI use, export controls, and national security requirements. This wide range of imperatives requires that boards ensure compliance programs evolve at the same pace as the legal environment.

AI systems can inadvertently create non-compliant outputs, automate processes that violate internal controls, or expose the company to penalties. Independent directors will need to review compliance readiness at every meeting, ensuring that legal and regulatory updates are incorporated into business processes before new rules take effect.

The rise of national-security-based restrictions on dual use technologies, AI systems, and quantum computing will require close coordination with counsel who understand the intersection of technology, trade policy, and foreign investment laws.

Strengthening the Audit Function in an AI-Driven Enterprise

AI creates both opportunities and risks for the audit process. Automated analytics can strengthen internal audit by improving anomaly detection and monitoring financial behavior, including by achieving testing and sampling rates of almost 100 percent. At the same time, AI systems introduce new audit challenges. Model drift, hallucinations, incorrect data ingestion, the potential for violations of technological controls, and unauthorized third-party tools can all undermine audit integrity.

Board members need to ensure that the internal audit function is equipped to evaluate AI systems, identify weaknesses, and confirm that financial reporting remains accurate. Auditors must assess whether AI-generated outputs are explainable, traceable, and aligned with corporate controls. Independent directors should also confirm that external auditors understand the company's AI architecture and can evaluate how AI is affecting financial statements, internal controls, and risk disclosures.

Communications and Stakeholder Trust

AI will also shape corporate communications and stakeholder expectations, and mishandled AI deployment can damage trust, erode customer confidence, alienate employees, and attract regulatory scrutiny. Boards need to support actions to reduce the risk of miscommunication magnifying reputational risk, particularly when AI systems malfunction or produce harmful outcomes. Corporate messaging also needs to accurately reflect how AI is used and what safeguards are in place.

Independent directors should ensure that disclosures are transparent and consistent across investor relations, customer communications, and internal messaging. In the future after AI, stakeholders expect clarity about how companies protect data, reduce bias, manage cybersecurity threats, and maintain ethical controls over AI systems. Communications strategies should reflect that expectation.

Reputational Risk and the New AI Accountability Standard

Reputational risk rises as AI becomes more integrated into decision-making. Failures involving AI systems are often perceived as governance failures. Boards need to oversee scenario planning for AI-related incidents, including algorithmic errors, employee misuse, adverse regulatory findings, and cyber intrusions involving AI models or datasets.

Companies will be judged on how responsibly they deploy AI and how quickly they address problems when they arise. Independent directors need to maintain oversight of AI ethics programs, third-party risks, and data governance rules. The public increasingly expects companies to demonstrate leadership in responsible AI practices. Boards that ignore this shift expose their organizations to strategic and reputational harm.

AI-Enabled State-Sponsored Cyberwarfare

The most serious emerging risk for boards is the rise of state-sponsored cyberwarfare powered by advanced AI. Adversarial nations are developing AI-driven tools that can penetrate corporate networks, map vulnerabilities, mimic employee behavior, and automate large-scale intrusions with unprecedented speed.

Boards need to assume that their companies are potential targets of AI-enhanced cyber operations conducted by China, Russia, Iran, and North Korea. Directors need to oversee investments in cyber resilience, including AI-based threat detection systems, strong identity controls, quantum-resistant encryption planning, and continuous monitoring of network behavior.

Cybersecurity should be treated as a central strategic risk rather than a technical function delegated entirely to the IT department. Boards must also prepare for scenarios involving destructive or disruptive cyber activity linked to geopolitical conflict.

Strategic Oversight and Future Readiness

The board's duty is to ensure that the company is prepared for long-term risks created by AI. This includes monitoring AI adoption across the enterprise, integrating AI considerations into capital allocation decisions, reviewing cyber and compliance posture, and ensuring that leadership teams communicate AI risks clearly to all stakeholders.

Independent directors need to cultivate expertise in cybersecurity, geopolitics, digital ethics, national security, and AI governance. The future after AI will reward boards that can see around corners, anticipate regulatory and technological shifts, and guide companies through intensifying complexities and risks.

AI is transforming corporate governance into a domain where strategic foresight is no longer optional. It is essential. Boards that embrace this responsibility will help their organizations remain resilient, trusted, and competitive in the years ahead.

CHAPTER 24

AI IMPLICATIONS FOR SPECIAL OPERATIONS FORCES

In the Future After AI, Special Operations Forces will face an operating environment transformed by the accelerating fusion of technology, economics, and national security. The traditional DIME framework, which describes the four elements of national power as Diplomacy, Information, Military, and Economics, still holds. However, AI is reshaping the balance among these four pillars in ways that create new vulnerabilities, new strengths, and new requirements for strategic adaptation.

The Threat to the Information Domain

The most profound shift is occurring within the Information component. AI threatens the integrity of information at a scale never before encountered. Synthetic data, deepfake leadership messages, algorithmic manipulation, large-scale bot amplification, and fully automated propaganda machines create an environment where truth becomes contested — and narrative control becomes fragile.

For Special Operations Forces, whose work depends heavily on trust, credibility, and the legitimacy of partners and populations, this represents a foundational challenge. The future SOF operator has to be as capable in the information domain as in kinetic or advisory missions because AI is transforming information into the most contested space of global competition.

Hybrid Conflict and the AI-Enabled Gray Zone

Hybrid conflict and gray zone engagements are likely to define the future operational landscape. Adversaries are blending cyber operations, economic coercion, political warfare, unconventional disruption, and targeted proxies to achieve their goals without triggering open war. AI accelerates each of these tools by giving adversaries greater reach, more rapid execution, and more adaptive influence strategies.

In this context, Special Operations Forces will be asked to operate at the intersection of human judgment and technological augmentation, where the ability to understand, leverage, disrupt, or counter AI becomes essential to mission success.

Economic Security as a Foundation for Irregular Warfare

One of the most significant transformations involves the connection between AI and economic security. Economic power and economic self sufficiency are becoming central to irregular warfare. Modern resistance movements, partner forces, and allied governments need economic foundations strong enough to withstand prolonged pressure from adversaries that increasingly rely on AI to manage economic influence, supply chain control, and resource denial.

AI allows both allies and adversaries to map supply chains in real time, identify critical choke points, expose weak links in industrial capacity, and understand financial vulnerabilities that could decide the outcome of long-term irregular conflict. For SOF, integrating economic security analysis into unconventional warfare planning is no longer optional. It is essential to ensure the endurance, resilience, and sustainability of resistance networks and partner forces.

AI as a Force Multiplier

AI functions as a force multiplier for Special Operations Forces. At the operational level, AI enhances route prediction, logistics optimization, partner force sustainment modeling, autonomous threat detection, and rapid mission rehearsal. At the strategic level, AI supports assessments of economic resilience, forecasts adversary supply chain behaviors, identifies pressure points that could degrade adversarial capabilities, and maps networks of gray zone influence. These tools give SOF the ability to operate more effectively in environments where technological speed and complexity increasingly determine outcomes.

Exploiting AI-Enabled Vulnerabilities in Adversary Networks

Adversaries who rely heavily on AI create new weaknesses. Centralized AI-controlled logistics become attractive targets. Large data centers become single points of failure. Automated decision tools can be deceived through model poisoning. High-tech supply chains reveal critical nodes that can be disrupted to limit adversary capacity to project power.

Special Operations Forces can exploit these vulnerabilities through cyber-enabled operations, disruption of automated logistics systems, interference with critical materials infrastructure, or collaboration with economic agencies to exploit adversary dependencies.

Building Partner Nation Resilience in the AI Era

Foreign Internal Defense and Security Force Assistance missions will also evolve. Partner forces in the Baltics, the Indo-Pacific, the Middle East, and other contested regions must be prepared to counter adversarial AI influence operations, protect their own information environments, and reduce dependence on adversarial economic networks. Enhancing partner nation resilience will increasingly depend on AI literacy, secure communications, diversified supply chains, and stronger economic foundations that allow allies to withstand coercion over months or years.

AI, Influence Operations, and Irregular Economic Warfare

AI has a powerful role in irregular influence operations. It allows for the automated identification of illicit funding networks, real-time tracking of adversarial capital flows, predictive modeling of trade coercion strategies, and rapid generation of counter-narratives that expose fabricated or manipulated information. SOF can apply these capabilities to degrade adversarial power, dismantle financial networks that support hostile regimes, and reinforce the economic and informational resilience of partner forces and resistance movements.

The Future of SOF in a DIME Environment Reshaped by AI

As the DIME framework evolves in the AI era, SOF must continue to integrate economic security into operational planning and strategic doctrine. This requires deeper collaboration with agencies such as the Department of the Treasury, the Department of Commerce, and the intelligence community.

It requires training that includes financial warfare, supply chain analysis, AI-driven influence operations, and the strategic use of economic tools to weaken adversarial capacity. It also requires wargaming that includes AI behaviors, economic coercion strategies, and irregular conflict scenarios that reflect modern technological realities, including AI.

Strategic Advantage in the Future Operating Environment

The central takeaway for SOF is that AI shifts the balance of national power. It threatens the Information domain more severely than any other part of the DIME framework, thereby magnifying the importance of the Information and Economic domains.

The Military and Diplomatic components will continue to matter. But their effectiveness will depend increasingly on whether the United States and its allies can preserve information integrity, maintain economic resilience, and counter adversaries who use AI to manipulate populations, weaponize supply chains, and exploit economic vulnerabilities.

AI is becoming the future operating system of countless industries and professions. SOF and the military more broadly have significant reasons for keeping humans in the loop for decision-making longer than almost any other profession. However, AI is still likely to become a fixture of the national security landscape and the bedrock of future activities.

Without rapid adaptation, adversaries will use AI to expand their influence faster than SOF can counter it. With deliberate integration of AI capabilities, economic security analysis, and information domain expertise, Special Operations Forces can preserve the strategic advantage necessary to succeed in irregular warfare, support partner nations, and defend United States interests in the long-term competition that defines the future international security environment.

CONCLUSION

NAVIGATING A NEW ERA OF OPPORTUNITY

We stand at the threshold of a new era — an era in which artificial intelligence becomes both invisible and indispensable.

The *Future After AI* is not simply a story about machines learning faster, working harder, or analyzing more. It is the story of how economies, societies, industries, governments, and individuals adapt to a world in which intelligence itself becomes a scalable resource, embedded into every system and quietly powering the global economy.

Throughout this book, I have shared one central idea: AI will be the bedrock of the future economy. Like electricity, semiconductors, cell phones, or the internet, AI will eventually fade from conversation because it will be everywhere. It will become the assumed operating system powering businesses and organizations in finance, technology, energy, healthcare, education, business, and national security.

But what will not fade — and what cannot be invisible — is the responsibility that leaders bear in this transition.

A New Economic Foundation

AI presents the opportunity to be one of the greatest productivity levers in economic history. It has the potential to add the functional equivalent of tens of millions of full-time workers to the U.S. economy and hundreds of millions globally. But realizing this upside requires judgment, discipline, and a culture of accuracy. AI can accelerate analysis, streamline workflows, and amplify human capability, but it cannot replace the need for creativity, relationships, trust, presence, and human connection.

A New Strategic Landscape

The future after AI is unfolding against the backdrop of Cold War Two, a contest defined as much by economic security and technological supremacy as by conventional geopolitical rivalry. The nations that dominate AI, semiconductors, energy, and critical minerals will shape the international order. The Technological Iron Curtain is falling. The information domain is under unprecedented assault. Economic power is national power. AI is not only a commercial tool. It is a strategic asset. It is also a strategic vulnerability. And leadership will determine which side of that ledger defines the future.

A New Imperative for Businesses and Boards

Corporate leaders face a world where governance, audit, compliance, communications, and cybersecurity risks are heightened by AI. Boards need to oversee AI deployment, ensure transparency, guide ethical use, and prepare for AI-enabled cyberwarfare.

In sales and marketing verticals, organizations need to prepare for the shift from B2B and B2C to B2AI and B2A, where AI agents and AI writ large become the focus of AI optimization strategies to drive top-line revenue.

Amid a bevy of AI-instigated disruptions and changes, the organizations that thrive will be those that pair technological adoption with strategic foresight and disciplined governance.

A New Role for National Security and SOF

Special Operations Forces and national security leaders need to adapt to hybrid conflict, AI-driven information warfare, economic coercion, synthetic influence, and rapid gray-zone escalation.

The DIME framework still applies, but AI has reshaped the balance by weakening the Information domain, magnifying the Economic domain, and demanding new capabilities from operators and allies. In the Future After AI, economic resilience is irregular warfare.

A New Opportunity for Urban Leaders

For urban leaders, it is crucial to act now by focusing on key strategies to foster perceived abundance, build strong cultures of data and education, embrace technology for smart city advantages, foster population health, and encourage social cohesion and stability.

Much will be asked of urban leaders in the future, but leveraging data and AI can help them deliver and improve the lives of their residents.

A New Responsibility for Individuals

For every professional, the future requires identifying your highest-value work, delegating everything else to AI, and doubling down on the skills that cannot be automated: judgment, trust, relationships, originality, and uniqueness. AI gives us more time. It is our job to invest that time where it can have the biggest material impact, especially by connecting with people, expanding our capabilities, focusing on doing the work only humans can do, and sharing authentic, human stories and insights.

Preparing for the Future After AI

The future is uncertain, but uncertainty does not prevent preparation. The ability to see around corners, weigh multiple futures, and think beyond the present gives leaders the advantage. Whether using the Four Futures, STEEP, or Almost Now / Maybe Someday frameworks, the path forward is clear: strategic foresight is no longer optional.

Above all, the best advice I give to executives is to make sure that their businesses enhance their ability to be on PAR — enhancing their *preparedness, adaptability, and resilience* — for more complex and complicated challenges posed by rapid change and AI technologies.

A Call to Lead in an Uncertain Era

The future will not wait, and AI is already reshaping the world. Hopefully, this book has made you more prepared — and inspired to capture upside opportunities for your organization and profession.

The future after AI will reward those who embrace innovation early, recognize its risks honestly, and act decisively. It will reward countries that secure energy supplies, restructure supply chains, and strengthen economic security. It will reward companies that invest in governance, resilience, and transparency. It will reward cities that focus on fostering environments of perceived abundance. And it will reward individuals who cultivate relationships, build trust, demonstrate original thought, and commit to lifelong learning.

This is not a moment for hesitation. AI is here. Soon, we will no longer speak of AI with wonder or curiosity as it becomes the new operating system of businesses, industries, and governments.

The question is not whether we are ready for AI. The question is whether we are ready to lead in the future *after* AI, when AI is the ubiquitous, assumed, and invisible operating system — and bedrock of our future economy.

Balancing current risks and opportunities is harder than ever.
But it is easier now than it ever will be in the future.

~ Jason Schenker
November 2025

ENDNOTES

Chapter 1

1 Google. (2025). *Google Trends: Worldwide search interest for "internet"*. Retrieved November 16, 2025, from https://trends.google.com.

2 Ibid.

3 Google. (2025). *Google Trends: Worldwide search interest for "computer"*. Retrieved November 16, 2025, from https://trends.google.com.

4. Google. (2025). *Google Trends: Worldwide search interest for "cell phone"*. Retrieved November 16, 2025, from https://trends.google.com.

5. Google. (2025). *Google Trends: Worldwide search interest for "computer"*. Retrieved November 16, 2025, from https://trends.google.com.

6. Google. (2025). *Google Trends: Worldwide search interest for "cell phone"*. Retrieved November 16, 2025, from https://trends.google.com.

7. Google. (2025). *Google Trends: Worldwide search interest for "AI"*. Retrieved November 16, 2025, from https://trends.google.com.

8. Google. (2025). *Google Trends: Worldwide search interest for "quantum computing"*. Retrieved November 16, 2025, from https://trends.google.com.

9. Google. (2025). *Google Trends: Worldwide search interest for "AI"*. Retrieved November 16, 2025, from https://trends.google.com.

10. Google. (2025). *Google Trends: Worldwide search interest for "quantum computing"*. Retrieved November 16, 2025, from https://trends.google.com.

Chapter 2

1. Steven Vaughan-Nichols. "Goodbye, Lotus 1-2-3: IBM Is Shutting the Doors on Lotus 1-2-3, the Software Program That Made the IBM PC and Microsoft Household Names." ZDNet, May 14, 2013. Retrieved November 27, 2025, from https://www.zdnet.com/article/goodbye-lotus-1-2-3/ .

2. Google. (2025). *Google Trends: Worldwide search interest for "Lotus 1-2-3"*. Retrieved November 16, 2025, from https://trends.google.com.

3. Ibid.

Chapter 3

1. Pew Research Center. (2025, September 17). *How Americans view AI and its impact on people and society: Views of AI's impact on society and human abilities* (by B. Kennedy, E. Yam, E. Kikuchi, I. Pula, & J. Fuentes). Retrieved on November 24, 2025, from https://www.pewresearch.org/science/2025/09/17/views-of-ais-impact-on-society-and-human-abilities/

2. Ibid.

3. The Futurist Institute.

Chapter 5

1. U.S. Bureau of Labor Statistics, Unemployment Rate [UNRATE], retrieved from FRED, Federal Reserve Bank of St. Louis; https://fred.stlouisfed.org/series/UNRATE, November 24, 2025.

2. Bureau of Labor Statistics. (2025, November 20). *The Employment Situation — September 2025* [News release]. Retrieved November 24, 2025, from https://www.bls.gov/news.release/empsit.nr0.htm

U.S. Employment and Training Administration, Continued Claims (Insured Unemployment) [CCSA], retrieved from FRED, Federal Reserve Bank of St. Louis; https://fred.stlouisfed.org/series/CCSA, November 24, 2025.

3. U.S. Bureau of Labor Statistics, Job Openings: Total Nonfarm [JTSJOL], retrieved from FRED, Federal Reserve Bank of St. Louis; https://fred.stlouisfed.org/series/JTSJOL, November 24, 2025.

ENDNOTES

4. Federal Reserve Bank of New York. *Unemployment rates for recent college graduates versus other groups: The labor market for recent college graduates.* Latest release August 1, 2025 (2025:Q2 data). Retrieved from https://www.newyorkfed.org/research/college-labor-market#--:explore:unemployment

5. U.S. Bureau of Labor Statistics, Civilian Labor Force Level [CLF16OV], retrieved from FRED, Federal Reserve Bank of St. Louis; https://fred.stlouisfed.org/series/CLF16OV, November 24, 2025.

6. U.S. Bureau of Labor Statistics, Job Openings: Total Nonfarm [JTSJOL], retrieved from FRED, Federal Reserve Bank of St. Louis; https://fred.stlouisfed.org/series/JTSJOL, November 24, 2025.

7. U.S. Bureau of Labor Statistics, Job Openings: Health Care and Social Assistance [JTS6200JOL], retrieved from FRED, Federal Reserve Bank of St. Louis; https://fred.stlouisfed.org/series/JTS6200JOL, November 24, 2025.

8. Federal Reserve Bank of New York. *Unemployment rates for recent college graduates versus other groups: The labor market for recent college graduates.* Latest release August 1, 2025 (2025:Q2 data). Retrieved from https://www.newyorkfed.org/research/college-labor-market#--:explore:unemployment

9. U.S. Bureau of Labor Statistics, Job Openings: Health Care and Social Assistance [JTS6200JOL], retrieved from FRED, Federal Reserve Bank of St. Louis; https://fred.stlouisfed.org/series/JTS6200JOL, November 24, 2025.

10. Federal Reserve Bank of New York. *Unemployment rates for recent college graduates versus other groups: The labor market for recent college graduates.* Latest release August 1, 2025 (2025:Q2 data). Retrieved from https://www.newyorkfed.org/research/college-labor-market#--:explore:unemployment

11. U.S. Bureau of Labor Statistics, Job Openings: Manufacturing [JTS3000JOL], retrieved from FRED, Federal Reserve Bank of St. Louis; https://fred.stlouisfed.org/series/JTS3000JOL, November 24, 2025.
U.S. Bureau of Labor Statistics, Job Openings: Transportation, Warehousing, and Utilities [JTU480099JOL], retrieved from FRED, Federal Reserve Bank of St. Louis; https://fred.stlouisfed.org/series/JTU480099JOL, November 24, 2025.

12. U.S. Bureau of Labor Statistics, Average Hourly Earnings of All Employees, Total Private [CES0500000003], retrieved from FRED, Federal Reserve Bank of St. Louis; https://fred.stlouisfed.org/series/CES0500000003, November 24, 2025.

13. Bureau of Labor Statistics, *Consumer Price Index — September 2025* (USDL-25-1502), released October 24, 2025, 8:30 a.m. ET. Retrieved from https://www.bls.gov/news.release/cpi.htm

14. U.S. Bureau of Labor Statistics, Job Openings: Transportation, Warehousing, and Utilities [JTU480099JOL], retrieved from FRED, Federal Reserve Bank of St. Louis; https://fred.stlouisfed.org/series/JTU480099JOL, November 24, 2025.

15. U.S. Bureau of Labor Statistics, Average Hourly Earnings of All Employees, Total Private [CES0500000003], retrieved from FRED, Federal Reserve Bank of St. Louis; https://fred.stlouisfed.org/series/CES0500000003, November 24, 2025.

Chapter 6

1. Bureau of Labor Statistics. (2025, November 20). *The Employment Situation — September 2025* [News release]. Retrieved November 24, 2025, from https://www.bls.gov/news.release/empsit.nr0.htm

2. “Bill Gates Explains the Internet to Dave” (1995, May 18). *Letterman*. YouTube video, 6:47. Posted by *CBS This Morning*, June 13, 2018. Retrieved on November 24, 2025, from https://www.youtube.com/watch?v=fs-YpQj88ew

3. US Debt Clock, *World Debt Clocks* (accessed November 24, 2025), https://www.usdebtclock.org/world-debt-clock.html

4. Social Security Administration, *Fast Facts & Figures About Social Security, 2025*, SSA Publication No. 13-11785, released September 2025. Retrieved on November 24, 2025, from https://www.ssa.gov/policy/docs/chartbooks/fast_facts/2025/fast_facts25.pdf

ENDNOTES

Chapter 9

1. Pew Research Center. (2025, September 17). *How Americans view AI and its impact on people and society: Views of AI's impact on society and human abilities* (by B. Kennedy, E. Yam, E. Kikuchi, I. Pula, & J. Fuentes). Retrieved on November 24, 2025, from https://www.pewresearch.org/science/2025/09/17/views-of-ais-impact-on-society-and-human-abilities/

Chapter 11

1. Graphite. (n.d.). *More articles are now created by AI than humans.* Retrieved November 17, 2025, from https://graphite.io/five-percent/more-articles-are-now-created-by-ai-than-humans.
2. Ibid.

Chapter 14

1. McKinsey & Company. *The data center balance: How US states can navigate the opportunities and challenges.* By Adam Barth, Chhavi Arora, Gayatri Shenai, Jesse Noffsinger, & Pankaj Sachdeva. August 2025. Retrieved on November 17, 2025, from https://www.mckinsey.com/industries/public-sector/our-insights/the-data-center-balance-how-us-states-can-navigate-the-opportunities-and-challenges. p. 4.
2. Ibid., p. 2.
3. Ibid., p. 7.

Chapter 18

1. World Bank. *Population Estimates and Projections* database, indicators SP.POP.TOTL, SP.URB.TOTL, SP.URB.TOTL.IN.ZS, and SP.URB.GROW. Retrieved September 23, 2025, from https://datacatalog.worldbank.org/search/dataset/0037950/World-Bank-Population-Estimates-and-Projections
2. Ibid.
3. Ibid.
4. Ibid.
5. The information in this section was provided to me directly as a member of the Seoul International Policy Advisory Group.

Chapter 20

1. Dator's Four Futures, Foresight University. Smart's Four Futures (adapted from Dator 1979), Foresight University. Retrieved on March 17, 2025, from https://foresightguide.com/dator-four-futures/
2. The Futurist Institute.

Chapter 21

1. Anderson, Benedict. *Imagined Communities: Reflections on the Origin and Spread of Nationalism.* Revised ed. New York: Verso, 1991.

PUBLISHED TITLES

Prestige Professional Publishing was founded in 2009 to produce insightful and timely professional reference books. We are registered with the Library of Congress, and we have published 37 books.

After the COVID Vaccine
Be the Shredder, Not the Shred
Cold War Two
Commodity Prices 101
Electing Recession
Financial Risk Management Fundamentals
Futureproof Supply Chain
A Gentle Introduction to Audit and Due Diligence
Jobs for Robots
Jobs for Robots: COVID Edition
Midterm Economics
Quantum: Computing Nouveau
Reading the Economic Tea Leaves
Recession-Proof Career Strategies After COVID
Robot-Proof Yourself
Spikes: Growth Hacking Leadership
Strategic Cost-Cutting
Strategic Cost-Cutting After COVID
The Author Institute Method
The Dumpster Fire Election
The Economics of Revolt and Revolution
The Fog of Data
The Future After AI
The Future After COVID
The Future of Agriculture
The Future of Energy
The Future of Finance is Now
The Future of Finance After COVID
The Future of Food and Agriculture After COVID
The Future of Travel
The Promise of Blockchain
The Robot and Automation Almanac - 2018
The Robot and Automation Almanac - 2019
The Robot and Automation Almanac - 2020
The Robot and Automation Almanac - 2021
The Robot and Automation Almanac - 2022
Write the Future

ABOUT THE AUTHOR

Jason Schenker is the President of Prestige Economics and Chairman of The Futurist Institute. He is a top-ranked economist, leading futurist, bestselling author, and in-demand keynote speaker who delivers high-impact presentations on economic trends, financial markets, geopolitical risks, AI, technology trends, and the future of work. Through his talks and research, Mr. Schenker helps clients anticipate risks and seize opportunities.

As a trusted voice on the economy, financial markets, geopolitics, and technology trends like AI and quantum, Mr. Schenker has delivered over 1,250 keynotes for public and private companies, corporate boards, industry groups, institutional investors, and government entities. He has spoken at events hosted by the U.S. Federal Reserve, the London Metal Exchange, the Singapore Exchange, the AFP, AICPA, SXSW, the CIA, the DOD, and NATO.

Mr. Schenker's speeches focus on the economy, financial markets, commodities, supply chain, futurist innovation and technology, artificial intelligence, blockchain, quantum computing, energy and sustainability, geopolitics, demographics, and other trends. He has been speaking and writing on AI, automation, and robots since 2016.

As a prolific thought leader, Mr. Schenker has created over 43 LinkedIn Learning courses on AI, economics, finance, and technology, which have been taken by more than 1.5 million students. He is the author of 37 books and 16 #1 bestsellers, including *Jobs for Robots, Futureproof Supply Chain, The Future of Finance is Now, Quantum: Computing Nouveau, The Promise of Blockchain, The Robot and Automation Almanac, The Future of Energy, Recession-Proof, The Future of Food and Agriculture After COVID, The Future After COVID*, and *Cold War Two.*

Bloomberg News has ranked Mr. Schenker the #1 forecaster in the world in 35 categories since 2011, including for his forecasts of crude oil prices, natural gas prices, gold prices, industrial metals prices (copper, aluminum, nickel, tin, zinc), major currency rates (the euro, the British pound, the Australian dollar), agricultural commodity prices (coffee, cotton, sugar, soybeans), U.S. new home sales, U.S. ISM Manufacturing, U.S. payrolls, and the U.S. unemployment rate. He is a Forbes contributor, LinkedIn Top Voice, former Bloomberg Opinion columnist, and media commentator on CNBC, CNN, the BBC, and Bloomberg, including as a guest host on Bloomberg Television.

Mr. Schenker created a rigorous program for The Futurist Institute that includes courses on The Future of Work, The Future of Transportation, The Future of Data, The Future of Finance, Futurist Fundamentals, The Future of Energy, The Future of Leadership, The Future of Healthcare, and The Future of Quantum Computing.

Mr. Schenker is one of the 100 CEOs on the non-partisan Texas Business Leadership Council, the Texas affiliate of the Business Roundtable that advises elected officials of both parties at the state and federal levels. He is a Board Leadership Fellow of the National Association of Corporate Directors and a Fellow of the Energy Institute. Mr. Schenker is also the AI and Emerging Technologies advisor on the Seoul International Policy Advisory Group.

Before founding Prestige Economics in 2009, Mr. Schenker was a Risk Specialist at McKinsey and Chief Energy and Commodity Economist at Wachovia (Wells Fargo). He holds master's degrees in economics, negotiation, and German, as well as a bachelor's in history and German. He has also been a Certified Financial Planner (CFP®) since 2010.

Mr. Schenker is currently Faculty for the Joint Special Operations University at U.S. Special Operations Command (USSOCOM). He is also an Adjunct Fellow at the Center for Strategic and International Studies and an advisor to the U.S. State Department and the Foreign Service Institute.

COLD WAR TWO

In *Cold War Two*, top-ranked forecaster and Chairman of The Futurist Institute, Jason Schenker, offers a futurist perspective on the geopolitical and geoeconomic risks, and what they mean for businesses, policymakers, investors, and national security leaders.

DISCLAIMER

FROM THE PUBLISHER AND THE AUTHOR

The following disclaimer applies to any content in this book:
This book is commentary intended for general information use only and is not investment advice. Neither Jason Schenker nor Prestige Professional Publishing, LLC makes recommendations on any specific or general investments, investment types, asset classes, non-regulated markets, specific equities, bonds, or other investment vehicles. Neither Jason Schenker nor Prestige Professional Publishing, LLC guarantees the completeness or accuracy of analyses and statements in this book, nor do Jason Schenker or Prestige Professional Publishing, LLC assume any liability for any losses that may result from the reliance by any person or entity on this information. Opinions, forecasts, and information are subject to change without notice. This book does not represent a solicitation or offer of financial or advisory services or products; this book is only market commentary intended and written for general information use only. This book does not constitute investment advice. All links were correct and active at the time this book was published. All content in this book is derived from open-source information and publicly available data; no classified or restricted material has been used in its analysis or presentation. All content in this book is unclassified.

Prestige Professional Publishing, LLC
4412 City Park Road #4
Austin, Texas 78730
www.prestigeprofessionalpublishing.com

THE FUTURE AFTER AI

Expectations for Artificial Intelligence as the New Operating System of Finance, Technology, Energy, Healthcare, Education, and Business

ISBN: 978-1-957772-40-0 *Kindle*
978-1-957772-37-0 *Paperback*
978-1-957772-24-0 *Hardback*

Made in the USA
Columbia, SC
05 December 2025

2cc64d43-c349-465b-a32a-2b21715340a3R01